Peach Orchard

A Gettysburg Battlefield Guided Tour

2020

Second Edition ISBN: 978-0-578-77199-1

Other books by Ralph Siegel

"Top Ten at Gettysburg"

(co-author with Rich Bellamy, Jay Jorgensen, Larry Korczyk,
Michael Rupert, William Schuber and James Woods;
Jay Jorgensen, editor)

GALE HOUSE PUBLISHING, TRENTON NJ

Peach Orchard: A Gettysburg Battlefield Guided Tour

Introduction

Welcome to this practical and user-friendly touring guide for walkers, hikers, drivers, cyclists, first-time visitors and armchair generals one and all. During the authors' 25 years combined experience in guiding at Gettysburg, we have noticed it is rare to see someone walking around the National Military Park with a guidebook open in their hands. On the occasions we have observed this, the visitors often are not following the guidebook as intended because most field guides and audio tours are inflexible and the tour stops are too lengthy. Some books with "guide" in the title are full-sized volumes more suited to an armchair than to a car seat or hiking pack.

Our set of guided tours fits every need to be tailored to visitors' interests on any given day.

The tour stops are numbered. Do not feel obligated to follow them. Start where you want, end where you want, walk, hike, drive, read a little, read a lot or stay home and read the whole thing legs-up in your favorite lounge chair. (And if you happen to catch us on the battlefield, throw the damn thing into the back seat; we'll give you the tour.)

The first, "Tour A: Walking" (page 1) is an easy stroll around the Peach Orchard itself, each of the four sides being about 100 yards.

The second, "Tour B: Hiking" (page 15) is the most involved: a four-mile field hike around this part of the battlefield for those who have good weather, the energy and the hours. It is in two parts so hikers can go halfway and finish the tour after lunch or on another day.

The third, "Tour C: Driving" (page 35) follows a 20-mile motor route. Since the National Park Service would really prefer you do not drive your car in between the peach trees, the Driving Tour covers the approaches to the Peach Orchard on July 2nd, 1863, including Lt. Gen. James Longstreet's infamous countermarch. The Driving Tour circles the field to show you how the Peach Orchard looks from important vantage points.

We have included a bonus "Cycling Tour" (page 71) with narratives from other tours but with its own set of directions. We have learned that unless provided with their own directions, cyclists would likely follow the Driving or Hiking tours, which is not good. The Driving Tour would involve doubling back on the same roads and the Hiking Tour could prompt cyclists to go off-road in the park, which is prohibited.

Finally, there is also a bonus "Couch Tour" (page 75) to direct the armchair expert's eye directly to the pages with the most provocative arguments and narratives. This guide can serve in this manner as a "stand-alone" read as well as an introduction to your tour, a tour companion or a post-tour follow-up.

☛ **ORIENTATION:** The Peach Orchard is near the southern end of the battlefield at the intersection of the Emmitsburg and Millerstown/Wheatfield roads. It is adjacent to National Park Service (NPS) Auto Tour Stop #10. It is easily detected by the row of four War Department tablets facing the intersection, by the many monuments and of course by the arching peach trees, which are more shapely and graceful than the gnarly apple orchards seen everywhere else in the park.

The four main streets in downtown Gettysburg are cardinal; that is, they run due north, south, east and west. But the Emmitsburg Road as it leaves town is actually heading *south-southwest* across the battlefield. However, for the purposes of these tours, we will consider the Emmitsburg Road to be *north* toward Gettysburg and *south* toward Emmitsburg.

To the west, the Millerstown Road snakes back and forth but ends up due west of the Peach Orchard at the point where it cuts through Warfield Ridge. To the east, however, the Wheatfield Road is actually heading *east-southeast* toward the Wheatfield (and beyond to the Taneytown Road.) For the purposes of these tours, we will consider the Wheatfield road to be *east*.

In 1863, the Wheatfield Road was an unnamed farm lane leading east and ending at the Rose Farm wheatfield. The national park name posted today, Wheatfield Road, is used for the tour guide. Likewise, Trostle Lane is the historical name but the present-day national park name, United States Avenue, is used for this guidebook.

Many esteemed volumes identify the line of woods on the horizon to the west and south of the Peach Orchard as a part of Seminary Ridge. This is the main Confederate line on July 2nd. The authors' interpretation of the local geography is that this wooded ridge is a distinct physical feature from Seminary Ridge and so we prefer to call it Warfield Ridge.

Abridged Preface to the forthcoming book "Peach Orchard: The Battle of Gettysburg Reconsidered."

The Peach Orchard is not a small story. Battlefield buffs and scholars know there are many more than four-score-and-seven volumes about the battle and about specific Gettysburg sites or events – from Willoughby Run to Culp's Hill to Little Round Top, most especially Little Round Top. Hundreds more books concern individual regiments or personalities. Academics sometimes call these monographs: Things with a small focus, small audience and small market.

Our topic may also seem narrow at first glance for those who might judge a book by its cover, but the authors consider the Peach Orchard story to be neither small nor narrow. This one spot on the battlefield seduces us into a reconsideration of the entire battle. The events of this book transpire mainly on July 2, 1863, a day which would end with what we believe to have been the bloodiest four-hours fighting of the American Civil War, really the bloodiest four hours of the American 19th Century.

The Peach Orchard was at center stage of the fighting on July 2nd. Therefore the Peach Orchard was central to the three days of battle, which also marked the 27th month, roughly the midpoint, of the 48-month war. Therefore the Peach Orchard can be seen as a central point of this ruinous, fratricidal conflict. Therefore the Peach Orchard can be seen as a central moment of the 19th Century in the Western Hemisphere. And so the Peach Orchard can be seen as a central event in the story of America.

Not such a small story at all.

Furthermore, the Peach Orchard story has so many interesting aspects.

One is that Gen. Robert E. Lee chose the Peach Orchard as the launching point for his largest attack of the battle, an attack which he ought to have cancelled in a battle he should not have fought.

Another aspect is that the actual fight that exploded around the Peach Orchard was a surprise. It was a surprise when Union Major Gen. Dan Sickles moved his Third Corps troops to occupy the Peach Orchard and the Emmitsburg Road. Sickles' move was a surprise to Confederate Gens. Lee and James Longstreet, who expected the Peach Orchard position to be open, inasmuch as the move was also a surprise to the Union army commander, Major Gen. George Gordon Meade. Sickles surprised everybody, driven as he was by a mix of ambition and confusion. Meade had a completely different idea of where he had intended the political general to place his six brigades.

Likewise, when Longstreet deployed his strike force south of the Peach Orchard along Warfield Ridge, it was a bloody surprise for Sickles and his men. It was this exact fact – that these two large forces moved into such proximity of one another without being aware of the presence of the other – that made the fighting on July 2nd so sudden and so violent.

Yet another interesting aspect about the fighting at the Peach Orchard involved artillery. Cannon usually take a secondary role in battle. On this day, however, artillery battalions and brigades fought each other alone near the Peach Orchard for almost an hour while the infantry brigades were busy moving into position. It was a close engagement of unprecedented ferocity. The numbers of guns involved compounded rapidly to the point where it became the largest direct-fire "gunfight" of the entire Civil War. Confederate Col. E. Porter Alexander, a great artillerist and an accomplished historian, would say: "I don't think there was ever in our war a hotter, sharper artillery afternoon than this."

From this artillery gunfight onward, the violence escalated. New Hampshire men faced off against and charged South Carolina men, only to change fronts to face a new attack from the Mississippi men. The exotic-looking Zouaves of Philadelphia stubbornly held the ground on the Joseph Sherfy farm against a frontal attack by one Mississippi regiment, only to find themselves flanked by two larger ones on their left and rear.

We know of no Gettysburg author who has disputed that the Peach Orchard was important to the battle. No significant work on the Battle of Gettysburg has overlooked it. But the authors here look at the fighting at the Peach Orchard with fresh emphasis, with the view that it was at this spot where victory and national independence came within the Confederate grasp, then slipped away.

The Peach Orchard. No point on the field is more critical.

Ralph Siegel
Mike Vallone
February 2017

Who they be?

The authors are battlefield guides licensed by the National Park Service. Licensed guides combine elements of the historian, the teacher and the stage performer with one more crucial ingredient: the battlefield itself. The battlefield, its grassy slopes and cracked boulders, its orchards and tree lines, woods and wetlands, roads surviving and lost, markers and monuments, placards and sculptures, its whispers and mists and memories, unify to form the indispensable stage. No guide can function without the stage and without the script, which was written by thousands of fighting men 150 years ago. The physical ground evokes and shapes the narrative. The complexities and enormity and horrors of the battle itself form what is also at its root a simple, single narrative to be told, rethought and "reconsidered."

Siegel and Vallone became friends and colleagues almost 20 years ago, overcoming the inter-borough blood feud that separates those from Queens and Brooklyn. They worked together in preparing for the National Park Service written examination in 2002 to become Licensed Battlefield Guides at Gettysburg. Both ranked in the top 20 of 120 test-takers, putting them on the list for the second-step oral exam. Mike passed in 2003 and was awarded his license while Ralph, who was much lower on the list, relied on Mike's help to secure his own license in 2004.

Mike had planned (threatened) for more than a decade to write a book about the Peach Orchard as an expression of his disruptive, creative ideas. He intended to upset the conventional understanding of the Battle of Gettysburg. Ralph cajoled Mike to begin the project for real in the summer of 2016. The following winter it was decided to produce this shorter "Battlefield Guided Tour" first yet to include in it the authors' most contentious points.

Who they owe?

Mike thanks battlefield friends and mentors Jim Clouse, Mike Phipps, Troy Harman, Garry Adelman, Tim Smith and Wayne Motts. Ralph thanks friends and mentors Andie Custer Donahue, Rich Kohr, Charlie Fennell, Rich Bellamy, Charlie Kingston, Hans Henzel and his friend among friends, John "California" Fitzpatrick. Both Mike and Ralph owe thanks and love to Harry Pfanz and Eric Campbell as teachers, to the irreplaceable John Heiser as a research guidance counselor and to our dearly beloved George Gargas from our same Class of 2002.

For this manuscript, the authors owe endless gratitude to LTC. Bill Hewitt (ret.) and Scott Slaymaker for their vigilant review and commentary. The highest thanks are owed John Fitzpatrick and Donna Leusner for their exhaustive line editing. Any mistakes or tangled sentences that made it into print are Ralph's fault. Any failings in historical judgment are entirely Mike's fault.

Ralph Siegel was born in Munich, Germany, raised in Queens and attended Rider University. After 25 years in the newspaper business, the last 10 with the Associated Press, he began a government career in land conservation finance. He conducts battlefield tours as a part-time weekender. He lives in Mercerville, N.J., with his wife, Donna Leusner. They have an adult daughter, Amy, and a rescue dog, Oliver.

Mike Vallone was born and raised in Brooklyn, attended St. John's University and worked in customer service for Bell/Verizon. He moved to Gettysburg in 2003 to become a full-time guide. He did groundbreaking work in developing tours for Hunterstown and Monterey Pass. Ill health forced him to give up guiding in 2014. He passed away on May 25, 2017, as this tour book was a work-in-progress. His ashes were spread in the Peach Orchard.

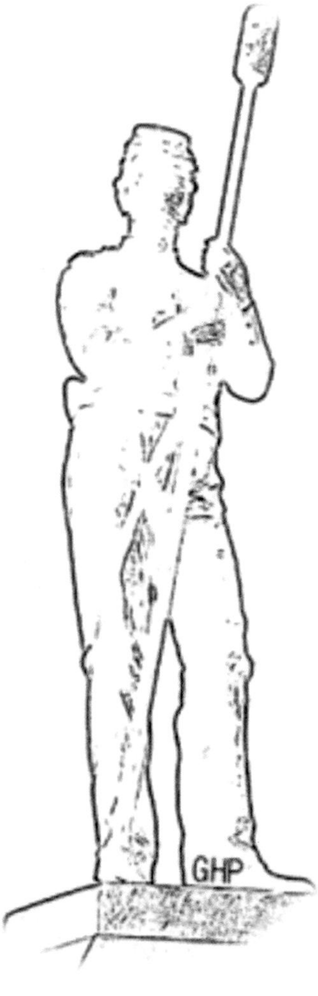
GHP

Tour A: Walking Tour

The Walking Tour covers the four corners and sides of the Peach Orchard with five tour stops. The total distance is only about 450 yards. The Walking Tour is conducted within the confines of the 3.5-acre orchard itself. There is no venturing along or across any roads so there are no safety concerns. Even the peaches are safe to eat.

Each Walking Tour Stop has a header followed by four finger-point bullets with specific instructions. These are followed by a "Quick Summary" with a few brief paragraphs and then "More Detail" with a narrative on the battle action.

- **GO:** This will provide instructions on where to walk.
- **ON THIS LEG OF THE WALK:** This will describe the significance of the segment of ground you will cross.
- **STOP:** This will tell you where to stop to take in your surroundings.
- **LOOK:** This will provide instructions on where to direct your attention so as to understand the narrative that follows.

Quick Summary: This will provide a few brief paragraphs to introduce and explain the corner where you have stopped.

More Detail: This will give a fuller battle narrative of the dramatic events occurring at each corner and edge of the Peach Orchard. It is intended for you to read this during the Walking Tour but it can also be read in advance or saved until afterward.

Walking Tour Stop #1:
Peach Orchard Salient Wayside Sign, Northeast Corner

- **GO:** Park the car in the south side pull-off on the Wheatfield Road. A nearby sign identifies this 110-yard stretch of asphalt as: "McGilvery Artillery Avenue." You can also park on Birney Avenue, the small lane that encircles the Peach Orchard on the south and east. Enter Birney Avenue from the Emmitsburg Road, drive down to the turn, follow the turn left, drive up and park anywhere on the right side with your car facing north toward the Wheatfield Road. There is a gravel strip so you can pull partly off of the road.

  (Never park on the Wheatfield Road itself under any circumstances. It is illegal and it is hazardous because this is a through road with two-way traffic and it crests at a blind hill. This is why the official NPS Auto Tour Stop 10 for the Peach Orchard is actually 150 yards back along Sickles Avenue.)

  Walk up the asphalt steps with handrails at the northeast corner of the Peach Orchard to bring you up to ground level.
- **STOP:** at the top of the asphalt steps.
- **LOOK:** toward the center of the Peach Orchard and toward the wayside sign directly in front of you.

Quick Summary: You can see here how the Peach Orchard is prominent high ground compared to the open, rolling fields that drop off in every direction, especially behind you. The orchard rows provide space for men and wagons yet the trees in 1863 were high enough to offer concealment. A natural feature like this being located at the intersection of two crucial roads enhances its importance.

> **What's a brigade?**
> A regiment is a unit of infantry of between 200 and 800 men formed in a specific city, county or region of a state. Three or more regiments form a brigade. Three to five brigades form a division. Two or three divisions form a corps.

On the first day of the battle, these two large armies had come to grips on the fields and ridges west and north of town – places you cannot see from here. At the climax of the fighting on July 1st, Confederate troops occupied the town itself while Union forces pulled back onto Cemetery Hill a mile and half north of you. By dawn of the second day, reinforcements that had been on the march through the night increased the size of both armies.

By the afternoon of July 2nd, the new Union line hinged on this Peach Orchard and extended north to your right for half a mile along the Emmitsburg Road. This line of troops faced west.

What is customarily described as a "salient" at the Peach Orchard had not yet formed. A "salient" refers to the corner angle that is created when a defensive line is bent back to face in two directions. The "elbow" point of such a bend can be a weak point. As first deployed, the Union troops of the Third Corps did not form a salient at all; in fact, their position was much worse. The brigades all faced directly west, leaving their left flanks unprotected. This was even more vulnerable than a salient.

More Detail: To your immediate right is a thick, plain obelisk monument commemorating the 141st Pennsylvania Infantry regiment. The name is on the other side. This will be discussed at the end.

> **What's a limber wagon?**
> A two wheeled-vehicle pulled by four or six horses. It carries a chest of ammunition and has a "pintle hook" on the back to tow the cannon, same as today's trailer hitch. "Limber up" is the command to hook up the guns and to get out.

Look at those delightful trees, how they form rows and are spaced apart from one another. Note how they create alleys with plenty of space for the movement of men and of limber wagons towing guns. The tree rows offer some concealment from threats on the horizon. Now imagine that the trees were tall enough to be

touching, forming canopies over these rows. This would offer even more concealment but of course not much in the way of actual cover from lead and iron.

Turn around and face away from the orchard to take in the landscape.

The minie-ball shaped monument to the 7th New Jersey Infantry is some 200 yards in front of you. Down the slope behind it is a barn with the famous cannonball hole on its brick face and with the smaller wooden-frame house behind trees to the right of the barn. This is the Trostle Farm. The large oak at the front right corner of the barn is a witness tree (here at the time of the battle). The tree marked the headquarters position for the Union Third Corps commander, Major Gen. Daniel E. Sickles.

> **What's a battery?**
> A unit of artillery with four to six guns. Each battery is organized into "sections" of two guns each. Batteries had from 75 to 120 men. In the Union army, four batteries or more would form a "brigade." On the Confederate side, three or four batteries formed a "battalion."

Half a mile beyond it, the dominant white dome of the Pennsylvania Memorial guides your eye to Cemetery Ridge and the Union battle line.

A glance farther to the right reveals the top edge of Little Round Top with monuments (and most likely crowds) clearly in view. A tree line cuts off your view of the base of the hill. This tree line forms the wooded western fringe of the Wheatfield.

The landscape integrity here today compared to its appearance in 1863 is good. It enables you to understand that the Peach Orchard is most definitely the high ground at a central position with open, sloping farm fields all around it.

White log Klingel farmhouse (left) and barn

Look left up the Emmitsburg Road north toward the town.

You can see how the road is generally at level with the Peach Orchard as it runs toward the white log house on the Emmitsburg Road. This white log house is the Klingel Farm. The landscape to the right (east) of the road clearly slopes down. This feature is sometimes referred to as the Emmitsburg Road Ridge.

Look farther left (west) across the Emmitsburg Road to Warfield Ridge in the distance.

That was the left end of the Confederate artillery position. Now look back at the Trostle Farm. That was the origin point from which the Union artillery batteries came into action, charging up the slope toward the Emmitsburg Road.

These two views enable you to understand that the Confederates to your far left – troops, gunners and commanders –
could not see over the Emmitsburg Road. They could not see the Trostle Farm at all and they could not see what forces might be deploying or coming up the slope from the farm to give them battle. Likewise, the Union men could not see what was waiting for them as they approached the Emmitsburg Road. They could not see what was happening or who might be forming battle lines along Warfield Ridge.

This obstructed view of the opposite ridges and battle lines influenced the movements of regiments and batteries. But it had a larger effect on army command decisions. Confederate commanders on Warfield Ridge could not see beyond the Peach Orchard or judge Union strengths, positions and movements. Likewise, Union commanders on Cemetery Ridge could not see beyond the Peach Orchard and so were unable to detect the deployment of enemy forces south along Warfield Ridge. From both perspectives, the Peach Orchard and Emmitsburg Road cut in half the commanders' views of the battlefield.

Turn back around to your original position facing into the orchard.

The wayside sign in front of you is entitled "The Peach Orchard Salient." It has a handy image of Brig. Gen. Joseph Kershaw. It explains the deployment of Sickles' corps followed by the attack of Kershaw's Brigade. The map in the right corner illustrates the Union lines in blue and Confederate line in red.

The painting by Edwin Forbes (uncredited) depicts the dense orchard to the upper right as Sickles, mounted in the center, rides boldly forward with his staff to inspect his lines. The white headquarters flag behind him bears the red diamond of the 1st Division, Third Corps. It is not known which regiment Forbes observed lying prone so far out toward the Rose Farm, probably the 3rd Michigan, but the artillery depicted shows the position of 1st New Jersey Battery B. Forbes faithfully depicts what is missing in so many artworks and movies: The thick crowd of horses that would have been standing behind every battery in action.

The authors have spent countless hours on the battlefield admiring and studying wayside signs like this one, especially when they were preparing for the licensed guide exam, which they took together in December 2002. We regard the waysides with affection and we are glad they are here. That said, it is always great fun to find errors. Make note of them and on your next visit share the errors with your friends – it makes you look smart.

This sign has three mistakes:

1) The Peach Orchard was owned by the Rev. Joseph Sherfy, the first Adams County farmer to be listed as a fruit grower, which today is the county's main industry. He would have been amused to see that the federal government thought his son John owned the orchard.
2) Brig. Gen. William Barksdale's Brigade shattered the Union lines in the Peach Orchard, then wheeled north to continue their attack. Brig. Gen. William Wofford's Brigade followed, swept through the Peach Orchard unopposed and continued on a separate line east. It is incorrect to imply Wofford's men participated in Barksdale's attack.
3) The "you-are-here" point shown on the map is most decidedly not where you are. It refers to the NPS Auto Tour Stop #10, which is about 150 yards behind you on Sickles Avenue. The wayside sign was probably designed to be installed back there at the official tour stop rather than here in the actual Peach Orchard.

Walking Tour Stop #2: "Twin Threes" 3rd Mich./3rd Maine Flank Markers, South Side

- **GO:** Turn left and walk south along the eastern edge of the Peach Orchard to the southeast corner and the 3rd Michigan monument. Examine the front of the monument, then turn left and continue 30 yards to the west to the pair of flank markers to give you a better vantage over the landscape.
- **ON THIS LEG OF THE WALK:** You are walking with the advance of two artillery units that moved to position on the edge of the orchard that you are approaching. You are also walking in the direction of the charge of the 2nd New Hampshire Infantry through the orchard to your right.
- **STOP:** at the pair of flank markers for the "Twin Threes" – the 3rd Michigan and the 3rd Maine.
- **LOOK:** south along the Emmitsburg Road, facing the Rose Farm.

<u>Quick Summary:</u> You have a commanding view of the Rose Farm in front of you and Warfield Ridge along the horizon to your front and right. This is the landscape across which the Confederate attack would begin on July 2nd.

The large square monument to your left, facing south, commemorates the 3rd Michigan Infantry regiment. It marks the end of the regiment's skirmish line, which continued down to the woods marking the Stony Hill.

Before the infantry attack began, a major artillery fight developed. The corner at your left was also the position occupied by six guns of 1st New Jersey Battery B. Four guns of the 15th New York Battery later deployed at your position. These 10 guns and others exchanged fire with Confederate batteries along Warfield Ridge in what would become a growing short-range artillery battle involving 86 guns on both sides. It was the largest such artillery "gunfight" of the entire Civil War.

<u>More Detail:</u> The 3rd Michigan monument provides a nice illustration of men in action amid the peach trees, which were depicted as being quite large. This bas-relief artwork is one of three on the battlefield done by noted Chicago sculptor Lorado Taft, although he is not credited. (The others by Taft are the 4th Michigan in the Wheatfield and the 5th Michigan on the Stony Hill, also uncredited.)

The original fieldstone house of the George Rose Farm and the ruins of the original stone barn are down the slope before you. Left of them is the cluster of woods that marks off the Stony Hill segment of the Wheatfield.

Like almost all monuments at Gettysburg, the 3rd Michigan to your left and the 3rd Maine monument to your right each has a pair of flank markers. The idea was that two flank markers are supposed to demark the edges of a battle line of a regiment with a monument in the center of it. You are standing by the right flank marker for the Michigan troops and the left flank marker of the Maine troops but neither is really correct. The 3rd Michigan's line actually extended all the way to the woods around the Stony Hill to the left of the Rose Farm. You can see that a line like this occupied by fewer than 300 men would have been long, thin and indefensible. It was also an odd position for a skirmish line. Skirmishers were usually deployed to the front of larger infantry forces. In this case, most parts of the skirmish line had no one behind them at all.

This threadbare Union line was a haphazard attempt to stitch a link between Col. Regis deTrobriand's brigade in the Wheatfield and Brig. Gen. Charles Graham's brigade here in the Peach Orchard. It is a good example of how the advanced line Sickles decided to occupy was too long for his brigades to hold in strength.

The six Parrott rifles of 1st New Jersey Battery B were arrayed down the slope to your left, down where the road is and beyond it across the front of the second orchard. These were the first Union guns to go into action here. Four Napoleons of the 15th New York Battery came in later to the right of the New Jersey guns, deploying exactly where you are standing.

What's a Parrott rifle?
A black iron gun that is rifled with "lans & grooves" inside the barrel to cause the bullet-shaped shell to spin when it is fired. The spin gives it greater range and accuracy. The gun tube is made out of cast iron with a wrought-iron collar welded around the "breech" at the back for strength.

What's a Napoleon?
A bronze smoothbore gun that fires a round shell or solid ball. Like the Statue of Liberty, the Napoleon when left outdoors turns green as it oxidizes. It was developed in France and is nicknamed for Emperor Napoleon III – not for the little Corsican fellow who conquered Europe.

There are no markers around here for these two batteries. The 15th New York Battery monument is far behind you back on the other side of the Wheatfield Road. The monument for 1st New Jersey Battery B is farther back still on Sickles Avenue about 150 yards behind the 15th New York monument. But your present position is where both of these batteries were actually fighting.

<u>Look to the right across the Emmitsburg Road.</u>

You can see the War Dept. observation tower in the middle of what the authors are calling the "Tower Woods." To the left of the Tower Woods, the ground opens up for about 500 yards to the white Phillip Snyder House. From the edge of the Tower Woods all the way to the Snyder house stood the line of Confederate Col. Henry Coalter Cabell's battalion of four batteries with 14 guns firing directly at you. As the artillery fighting began in earnest, 1st New Jersey Battery B traded fire with those guns on Cabell's line. The 15th New York Battery eventually came up to lend its weight to the fight.

What's canister?
A type of ammunition with dozens of large pellets crammed into a tin can. When fired it functions like a giant shotgun shell, blasting a spray of deadly pellets for a few hundred yards. It is terrifying in close action against troops.

With New York and Rhode Island batteries coming into action behind you and New York guns on the Devils Den, all of them facing off at short range against 48 Confederate guns, this became the largest artillery "gunfight" of its kind in the Civil War.

<u>Face back south.</u>

The first Confederate infantry units you would be able to see advancing on July 2nd were the four regiments of the Texas Brigade on Warfield Ridge straddling the Emmitsburg Road gap. They would march from right to left across the Michael Bushman Farm, aiming for the Devil's Den. The summit of Big Round Top is visible southeast of you and the Devil's Den is directly beneath it but concealed from view by the woods. As the Texans began their advance, the New Jersey and New York batteries were presented with an extraordinary field of fire into the left flank of the Texas line. The Confederate division commander, Major. Gen. John Bell Hood, rode into battle at the head of his old familiar Texans – and he did not get far. Somewhere on that slope Hood was crippled when fragments from an exploding shell tore up his left shoulder. While the wounding of Hood is often attributed to the fire from a Union battery near Devil's Den, the guns of 1st New Jersey Battery B right next to you had a superior angle to shoot at the Texans' battle line. It can never be known but it is more plausible that a New Jersey shell took Hood out of action, condemning his four brigades to confusion and disarray for the remainder of the fighting on July 2nd.

Hood

<u>Look to the right again at the Tower Woods.</u>

Kershaw advanced his brigade out of those woods and reached the Emmitsburg Road, heading down the lane that leads to the Rose Farm. The New Jersey and New York guns switched their aim and eventually their ammunition to canister to fire on this new infantry force as it came closer. Kershaw had three units of his brigade under Col. John Kennedy turn left and move directly against the Peach Orchard. These gunners quickened the pace of their fire. One crewman proclaimed, "The men at the guns fairly flew to their work. The guns themselves seemed almost full of life, how they roared and thundered."

Walking Tour Stop #3:
2nd New Hampshire Monument, Southwest Corner

- **GO:** Turn right and walk west 65 yards along the rest of the southern edge of the Peach Orchard.
- **ON THIS LEG OF THE WALK:** You are passing along the combat line of the 2nd New Hampshire Infantry, which charged to this position to repulse the attack of South Carolina troops coming up from your left.
- **STOP:** at the painted post-and-board fence by the southwest corner where Birney Avenue intersects the Emmitsburg Road.
- **LOOK:** south along the Emmitsburg Road, facing the gap in the Warfield Ridge.

Quick Summary: The men of Brig. Gen. Joseph Kershaw's South Carolina Brigade began their advance from Warfield Ridge to your right, crossing the Emmitsburg Road toward the Rose Farm. Once across the road, a detachment of 800 men – two regiments plus one battalion – broke off and turned toward the Peach Orchard and Wheatfield Road.

The South Carolina attack moved against the position where you are now standing and against the south edge of the orchard along your left. The 2nd New Hampshire and 3rd Maine regiments came up from behind you, formed ranks on the south side of the Peach Orchard and repulsed part of the attack, driving the Confederates down the slopes and back up to the Rose Farm lane. The pyramid-shaped monument to the 2nd New Hampshire is right next to you. The thin obelisk marker to your right indicates this was the advanced position of the 68th Pennsylvania regiment, which held the line along the edge of the orchard that faces west.

More Detail: On your walk you passed the regimental monument to the 3rd Maine, an obelisk with the diamond balanced on the tip. That diamond is the "corps badge" emblem of Third Corps. (Every Union regimental monument at Gettysburg has the corps badge emblem somewhere on it.) To your right, shielded by the small segment of post-and-board fence, is a weathered forward marker for the 68th Pennsylvania. Their primary monument is on the Wheatfield Road about 100 yards behind you. The marble obelisk next to you represents their advanced position in the orchard and according to the inscription is also intended as a tribute to their dead.

Look left to the Rose Farm.

Behind the farm and to the right of the Stony Hill woods you should be able to detect a gap in the forest that provided access back to the Rose Farm wheat field.

> **What's a corps badge?**
> An emblem Union troops wore on their hats to designate their corps. When veterans returned to place monuments on the battlefield in the 1880s, they put their corps badges on them. The Third Corps badge was a diamond, Second Corps a trefoil or poker "club," Fifth Corps a Maltese cross.
>
> ● **I CORPS**
> ♣ **II CORPS**
> ◆ **III CORPS**
> ✠ **V CORPS**
> ✚ **VI CORPS**
> ☾ **XI CORPS**
> ★ **XII CORPS**

Look south down the Emmitsburg Road.

Kennedy's attack comprised the 8th South Carolina infantry in front of you and the 3rd South Carolina infantry battalion to the left of them. As the men from the Old South State moved down the slope from the Rose Farm lane, the Union batteries to the left raked them with merciless fire. It is likely the guns farther east along the Wheatfield Road under Lt. Col. Freeman McGilvery also turned their barrels and joined in the fire, though they would have had to be careful not to clip the New Jersey gun crews. To the left of the 3rd battalion, Kennedy's own 2nd South Carolina infantry moved directly at McGilvery's guns on the Wheatfield Road.

Bailey

At the bottom of the slope before you, Kennedy's men would have crossed the marshy ditch at the lowest point and then started their advance up toward the Peach Orchard. This ditch is the trickling headwater of West Branch Plum Run. (It is sometimes called Rose Run.) It flows east across the Rose Farm and merges into a larger branch of Plum Run south of Devil's Den.

(While the authors have great praise for the National Park Service (NPS) work to restore the battlefield, this is one spot where the effort has slipped. A row of woody shrubs and trees has been allowed to form a barrier in the West Branch Plum Run ditch. This had been opened up a decade ago and was open and clear in 1863, but unfortunately the thicket you see has been allowed to fill in and now it blocks your view. These bushes and trees would have interrupted the advance of Kennedy's men and simultaneously provided much-needed cover – except it was not there.)

Col. Edward Bailey's 2nd New Hampshire Infantry had concealed itself in the orchard during the heavy artillery fire. The men would be lying down amid the trees behind you. But when the Confederates came up the slope and into the Peach Orchard, the New Hampshire men formed ranks and came charging through the trees to deal with Kennedy's advancing troops. The 2nd New Hampshire's right flank extended across the Emmitsburg Road and their advance would have swept right over your location.

The collision of these two forces is all around you. The South Carolinians may have had better numbers but the Granite State warriors had gravity and ferocity on their side. They drove the South Carolinians out of the Peach Orchard and back down the slope to the Plum Run ravine with heavy loss. Union officers had to ride after their men to halt the charge, one of them crying out: "Stop them. They are going all the way to Richmond." Instead, Bailey's men formed a firing line along the fence in front of you, which may have been 10 or 20 yards farther down the slope in 1863.

As the Granite State men kept up their fire, the men of the 3rd South Carolina battalion and 2nd South Carolina infantry advanced from the Rose Farm to your left. The men of the 3rd Maine infantry charged through their section of the orchard to confront part of this attack, coming into line to the left of the 2nd New Hampshire. The New Hampshire men probably lent the weight of their rifle fire into thwarting the 3rd battalion but they stayed in their Peach Orchard position.

It was during this ill-fated part of the attack by Kennedy's men that a horrible mistake in orders occurred, causing the South Carolinians to halt their advance toward the Wheatfield Road, shift by their right flank and march east toward the woods on the fringe of the Wheatfield. Moments earlier, McGilvery's gunners had been preparing to pull out, afraid they were about to be overrun at a position where they lacked infantry support. Now the two batteries blasted Kennedy's line at close range with canister as the Confederates inexplicably turned to present their flank to the muzzles of the cannon.

Non-historic hedgerow below Rose Farm lane

Walking Tour Stop #4:
1st New York Battery G, Northwest Corner

- **GO:** Turn around and walk north along the western edge of the Peach Orchard. The Emmitsburg Road will be on your left as you walk.
- **ON THIS LEG OF THE WALK:** You are walking across the muzzles of infantry rifles, first of the 68th Pennsylvania infantry, then of the 2nd New Hampshire. You are also passing in front of cannon, first the bronze Napoleon guns of 1st New York Battery G positioned amid the trees, then the iron Ordnance Rifles of 1st Pennsylvania Battery C&F.
- **STOP:** at the pair of green bronze Napoleon cannon and the monument not far from the northwest corner where the Emmitsburg Road intersects the Wheatfield Road.
- **LOOK:** west along the Millerstown Road toward the two farmhouses on Warfield Ridge.

Quick Summary: This monument marks part of the position of 1st New York Battery G exchanging fire with a much stronger line of Confederate batteries along Warfield Ridge 550 yards in front of you.

Brig. Gen. William Barksdale's Mississippi Brigade was behind those Confederate guns in the area of the two farmhouses you can see on the Millerstown Road and in the woods to their right. While Kershaw's men failed to break the Union hold on the Peach Orchard, the charge of Barksdale's Brigade would smash it. By the time that attack came, the New Yorkers' Napoleon guns had been replaced by the rifles of 1st Pennsylvania Battery C & F. (That battery's twin monuments are behind you to your right.)

The 21st Mississippi infantry regiment would march straight into the Peach Orchard in line of battle over the point where you are standing. The 2nd New Hampshire infantry would change position to face this attack.

More Detail: 1st New York Battery G was commanded by Capt. Nelson Ames. Looking out along the Millerstown Road are two farms about 500 yards away. They are the Joseph Warfield Farm on the left side of the road and the Christian Shefferer Farm across the street on the right side of the road. Ames' left-most section was closer to Stop #3, firing south or southwest. The guns where you are standing would have likely been firing west or northwest to engage Confederate batteries posted on the open ground to the right of the Shefferer Farm.

If the light is good, those of you with good eyes will be able to pick out the Louisiana Memorial to the right of the Christian Shefferer house. It marks the ground that was occupied by four howitzers of the Branch Artillery of South Carolina, commanded by Lt. Stephen Capers Gilbert.

> **What's a howitzer?**
> A smaller, smoothbore bronze gun with short range that fires its round shells on a high arc. The Union Army of the Potomac had few but they were common in the ranks of the Confederate artillery.

Barksdale's Mississippi Brigade was in position around and behind those Confederate batteries. Ample accounts document that Barksdale was anxious, perhaps to the point of ill-temper, to unleash his four regiments in order to silence the Union guns in the Peach Orchard that were slamming shells into his men. No less than Lt. Gen. James Longstreet was there in person to hold Barksdale back with the admonition: "We are all going in presently."

Look over your right shoulder.

You will see across the Wheatfield Road the monument to the 68th Pennsylvania, known as the "Scott Legion." You are on the ground they were defending, a line from your position right across the Wheatfield Road and beyond through the farmyard of the John Wentz Farm, marked today only by the rectangular remnant of a stone foundation.

North across the Emmitsburg Road is the large red barn and, beyond it, the farmhouse of the Sherfy Farm. That was the position of the 114th Pennsylvania, known as "Collis' Zouaves" for their colorful Algerian uniforms of baggy red trousers, blue jackets and red stocking caps. They covered the right flank of the 68th Pennsylvania and the 68th covered their left. Also within the farm complex and the Sherfy House itself were the men of the 57th Pennsylvania infantry.

The Union infantry line from your position to the Sherfy House is the line Barksdale's men would hit once their powerful attack began. By that time, Ames' battery had withdrawn to the ridge behind the Trostle Farm. McGilvery's reserve battery, 1st Pennsylvania Battery C & F commanded by Capt. James Thompson, would have come into the Peach Orchard and Wentz Farm in their place – just in time to be overrun by Confederates.

Turn around and look into the Peach Orchard.

To your left on the northern edge of the orchard are the twin monuments for Battery C & F, each with a pair of guns facing into the trees. They were two separate batteries until May 1863. After the war their veterans would return to dedicate separate monuments. Battery C is the low monument with a granite carving of a cannon. The Battery F monument is more memorable with a beautiful statue of the rammer facing west into the sunset. It may seem silly to have the cannon that accompany these monuments facing south as if to fire directly into the trees. It would be even worse to have them facing north against their own troops. In truth, four of Thompson's guns were on the ground where you walked and are now standing. Two others were deployed across the street on the Wentz property, facing west.

Turn back around to face west across the Emmitsburg Road.

Barksdale's largest regiment, led by his senior and most trusted commander, Col. Benjamin Humphreys, was directly in front of you in the area of the farm houses. Because the Millerstown Road snakes back and forth, some of Humphrey's men would actually cross it twice. The 21st Mississippi battle line would march straight into the Peach Orchard over the point where you are standing. From your perspective, the 17th and 13th Mississippi were to the right of the 21st Mississippi. These regiments would not come into the orchard itself but into the position of the 68th Pennsylvania and the Wentz Farm to your right. At the far right end of the Mississippi brigade was the 18th Infantry, which would attack directly into the Sherfy Farm against the 114th and 57th Pennsylvania.

As the Mississippi wave began across these fields, the 2nd New Hampshire on the southern edge of the Peach Orchard saw the peril coming from their right. They executed a complex maneuver that only well-drilled veterans could have done, especially in the middle of an orchard being pelted by bullets and artillery shells. They turned about-face, wheeled to the right by companies, then turned about-face again to form a new battle line facing the Mississippi troops. One company captain, quite proud of this movement, boasted about it to another officer. Before a compliment could be rendered, a Mississippi bullet hit the captain in the head and killed him.

The Union men put up a hard fight but the line dissolved, the 68th Pennsylvania giving way first. The 2nd New Hampshire, isolated in the Peach Orchard, could not hold out long as the three Mississippi regiments shattered the Union line in the Peach Orchard and Wentz Farm, jeopardizing the Third Corps' entire advanced position. Brig. Gen. Charles Graham, wounded and his horse killed, would hit the ground and be made a prisoner. As Mississippi men poured into the orchard, Thompson's men would struggle mightily to get just one gun out. Capt. Patrick Hart's Irishmen in the 15th New York Battery would go to ground, letting the Mississippi men pass over them before getting up to pull out their guns. Such desperate heroics would earn one man from each of these batteries the Medal of Honor.

Look right toward the north.

In the fields beyond the Wentz Farm and fences is a monument out by itself with a pair of bronze statues side by side. If you get a closer look at this sculpture will notice it depicts a soldier clasping hands with a fireman. This is the monument to the "2nd Fire Zouaves," the 73rd New York infantry, mustered into service from among New York City firefighters. (It has become a popular subject for photographers since 9/11). The monument is far in advance of the other markers or monuments for the New York City-based Excelsior Brigade. Why? This supposedly marks the spot where the hard fighters of the 73rd charged to recapture a gun of Thompson's Battery.

Walking Tour Stop #5 (Final): 141st Pennsylvania Monument, Northeast Corner

- **GO:** Turn around and walk east along the Wheatfield Road edge of the Peach Orchard past the monuments for Battery C and Battery F.
- **ON THIS LEG OF THE WALK:** You are walking with the attack of the 21st Mississippi Infantry as it struck and broke the ranks of the 2nd New Hampshire, collapsing the Peach Orchard defense. Somewhere to your right you will pass the fallen Gen. Graham, wounded and soon to be made a prisoner.
- **STOP:** at the thick, plain obelisk monument at the northeast corner next to the asphalt stairs where you began the Walking Tour.
- **LOOK:** toward the east and down the recessed Birney Avenue lane to your right front.

Quick Summary: The recessed segment of the Birney Avenue lane in front of you is the final position of the 141st Pennsylvania regiment commanded by Col. Henry Madill. As the Mississippi troops came plowing through the trees behind you, the Pennsylvania men opened fire and brought their advance to a cold stop. This won precious minutes for the artillery crews in the orchard and on the Wheatfield Road to get their guns out.

The minutes were paid in blood as the 141st Pennsylvania was almost wiped out. No Union Third Corps regiment suffered a heavier percentage of losses (71%) at Gettysburg. The 141st Pennsylvania monument is to your left and it faces toward the north.

Madill

After breaking the Peach Orchard line, three of Barksdale's regiments wheeled to the north and pressed their attack to your left against the forces on the Emmitsburg Road. However, the 21st Mississippi kept moving forward through your position and down the slopes toward the Trostle Farm.

More Detail: Turn around and face back into the orchard.

You will again notice that the orchard is not flat. Moving east to west, the land rises slightly into the center, then declines back down toward the Emmitsburg Road. If you take a knee, you can no longer see the two Napoleon cannon next to the 1st New York Battery G monument at Stop #4. This slight rise of ground and the landscape around you, from Little Round Top and the Devil's Den through the Wheatfield and across the fields below you to the Peach Orchard itself, is the amphitheater of the bloodiest four-hours fighting of the American Civil War. And on this gruesome day, no Third Corps regiment – none – suffers a heavier percentage of losses than the 141st Pennsylvania. The regiment was from Bradford and Susquehanna counties on the northern border of the state but their commander, Madill, happened to be born not far from here in the village of Hunterstown just north of Gettysburg.

The slope of the ground within the orchard and the peach trees themselves will conceal for a time the advancing line of the 21st Mississippi regiment. Conversely, it will conceal a second Union battle line formed behind you by the men of the 141st Pennsylvania regiment.

Examine the monument to your right. The front faces the Wheatfield Road.

Perhaps you will agree it is rather ordinary and nondescript. On a typical battlefield tour by car or bus it might seem to be one of those monuments that is easily missed.

It should not be.

Turn around and face east again, looking away from the Peach Orchard.

Below you and to your right is the recessed segment of the Birney Avenue lane. There may have been a farm lane here, perhaps not, but there was some recess of ground between the Peach Orchard and the second orchard that belonged to the Rose Farm. This is the position where Madill placed many of his men in line of battle facing into the Peach Orchard. Sensing peril, one of his officers questioned if they had not better get out of there but Madill said he had no such orders. And as the big Mississippi regiment came plowing through the trees, eager to capture Union guns as a prize, the Pennsylvania men opened fire and stopped their advance. This bought precious minutes for the men of the 15th New York Battery and 1st New Jersey Battery B, and for McGilvery's batteries farther east down the Wheatfield Road, to limber up with what horses were still alive and to get out.

The men of the 141st Pennsylvania made an unimaginable sacrifice. The first volley by the 21st Mississippi shattered the Union regiment's line. The regiment was small, just 209 men, and the deadly firefight was brief. The 141st would lose 149 men killed, wounded and missing. With the entire color guard down, Madill, his horse shot out from under him, would carry the flag himself and lead some 20 survivors down the slope toward the Trostle Farm and Sickles' headquarters. (Madill would locate 30 more survivors later.) Sickles himself would ride out to meet Madill and ask him if he could not hold. Madill in tears would reply, "With what men?"

Humphreys

At this point, Confederate Col. Humphreys came to where you are and saw a landscape crowded with scattered bands of Union troops on the run and the gun crews of the 5th and 9th Massachusetts batteries firing their weapons on the Wheatfield Road or trying to limber up and get out.

Look left toward north up the Emmitsburg Road.

Col. Humphreys would see the other regiments of the Mississippi brigade wheeling in that direction and reforming to advance their battle line northward. Perhaps Humphreys would even spot the fiery Barksdale waving his hat frantically and urging his men forward. The aroused Mississippians would flank and turn the Union regiments out of the Sherfy Farm on the other side of the Emmitsburg Road and then strike the next line of Union regiments forming behind the white-log Klingel Farm.

Humphreys knew the plan had been for him to turn his regiment and join in this second phase of Barksdale's vainglorious charge. But Humphreys was a combat veteran and now the battlefield conditions compelled a change in tactics.

Look back east down the slope.

Visualize the commotion. Imagine what decision you might have made. No competent commander would pivot his regiment left to the north when doing so would leave this congregation of Union troops and guns in his rear, no matter how disorganized the Billy Yanks might appear at the moment.

Col. Humphreys held the line of attack that had brought him through the Peach Orchard. He did not turn. He advanced his regiment straight ahead down the Wheatfield Road and down the slope toward the Trostle Farm. He had the numbers and he had the momentum to drive those Union forces and to sweep the battlefield. Barksdale, one of Mississippi's "fire-eater" radicals, would simply have to make do with his other three regiments as he turned north to roll up the Union line and win the war all by himself.

McGilvery

Now picture the chaotic scene of retreating troops and batteries limbering up as the lone sea captain, McGilvery, rides back and forth across the field barking orders and directing men to form a second line of guns that would put a bloody stop to the surging Mississippi regiment.

Turn around again and look back west toward Warfield Ridge.

As the bloodshed on July 2nd neared its climax, one more dramatic event unfolded in the Peach Orchard as

Col. E. Porter Alexander advanced his battalion of six batteries in an artillery "charge." It was a rare event in the Civil War because it could only be executed by experienced troops in the hands of skilled battery officers. All of his 24 guns limbered up together, charged some 600 yards across the ground, then unlimbered and rapidly began firing from new positions in the Peach Orchard and to your right (north) across "Excelsior Field" almost all the way to United States Avenue.

Alexander's gunners took advantage of an elevated position to rain hellfire on clusters of blue troops wherever they could be seen. One shot punched the prominent hole that appears to this day in the brick facing of the Trostle Barn. Another shot shredded Sickle's leg, which appears to this day at the Army's Museum of Health & Medicine in Silver Spring, Md. Yet another killed Col. George Willard, whose brigade brought a final stop to Barksdale's charge – and whose men would riddle Barksdale with bullets.

Alexander's fire would not go unanswered. McGilvery's guns exchanged fire with the Confederate guns until well after dark, the gunners taking aim at the red sparks of their opponents' muzzles. Alexander said, "My guns stood and fired until it was too dark to see anything more and both sides were glad to stop and rest."

Statue on the monument for 1st Pennsylvania Battery F within the rows of trees

MONUMENT PHOTOS without captions

PAGE 2 141st Pennsylvania
PAGE 3: 7th New Jersey
PAGE 5: 3RD Michigan
PAGE 7 2nd New Hampshire (top) & 3rd Maine
PAGE 9 1st Pa Battery F & Louisiana Memorial
PAGE 10 1st Pa Battery C & 73rd New York
PAGE 12 141st Pennsylvania

GHP

Tour B: Hiking Tour

The Hiking Tour covers four miles with 15 tour stops, bringing you around and through the amphitheater of the bloodiest four hours fighting of the American Civil War. It is split into two parts of about two miles each.

The Hiking Tour, like the Walking Tour, will begin at the Peach Orchard itself and relies on many of the same narratives. The Hiking Tour then ventures out to critical points and lines on the ground where all forces converged.

The Hiking Tour will traverse the Confederate infantry and artillery lines. It will take you across the Union positions along the Emmitsburg Road, down to the Trostle Farm and to the Plum Run Line. Then the Hiking Tour will cover the ground of the original Union position with important vantages to better appreciate the importance of the Peach Orchard.

BEFORE YOU GO: Please make note that extra caution is warranted for the Hiking Tour from Stops #3 to #5 and Stops #8 to #10 because they follow busy Emmitsburg Road in an area where the speed limit is 40 mph. Always use the left shoulder of the Emmitsburg Road so you face oncoming traffic. Always use caution when crossing the Emmitsburg Road. It is common for motorists to exceed the speed limit in this section. NEVER assume motorists can see you or will slow down to accommodate you as you cross the road. Bear in mind most drivers are using this as a "through" route and are not visiting the park.

Likewise, caution is warranted for the Hiking Tour between Stops #7 and #8 and again between Stops #13 and #15 because they follow the Millerstown and Wheatfield Roads. These are less busy than the Emmitsburg Road and have a lower speed limit but they are also public through roads with two-way traffic. On both roads the best shoulders happen to be on the right side in the direction you will be walking.

Each Hiking Tour Stop has a header followed by four finger-point bullets with specific instructions.

- **GO:** This will provide instructions on the route to the next Hiking Tour Stop.
- **ON THIS LEG OF THE HIKE:** This narrative will describe the significance of the ground you will cover.
- **STOP:** This will provide specific instructions on where to stop and read the narratives.
- **LOOK:** This will provide instructions on where to direct your attention to understand the narrative that follows.

Quick Summary: This will provide a few brief paragraphs to introduce and to explain the significance of the place where you have stopped.

More Detail: This will give a fuller battle narrative of the dramatic events occurring at each Hiking Tour Stop. In a few cases the More Detail narrative will refer back to the Walking Tour because the narrative at that specific spot is the same.

Hiking Tour Part I – Stop #1: Peach Orchard Salient Wayside Sign

- **GO:** Park on Sickles Avenue at NPS Tour Stop #10. You may also park on Birney Avenue, the small lane that encircles the Peach Orchard on the south and east. Enter Birney Avenue from the Emmitsburg Road, drive down to the turn, follow the turn left, drive up and park anywhere on the right side with your car facing north toward the Wheatfield Road. There is a gravel strip so you can pull partly off of the road. There is also the pull-off on the south side of Wheatfield Road but you should be polite and leave that for quick-stop visitors and buses since your car is going to be there for a few hours.
  (Never park on the Wheatfield Road itself under any circumstances. It is illegal and it is hazardous because this is a through road open 24 hours to two-way traffic and it crests at a blind hill.)
  Walk up the asphalt steps with handrails at the northeast corner of the Peach Orchard to bring you up to ground level.
- **STOP:** at the top of the asphalt steps.
- **LOOK:** toward the center of the Peach Orchard and toward the wayside sign directly in front of you.

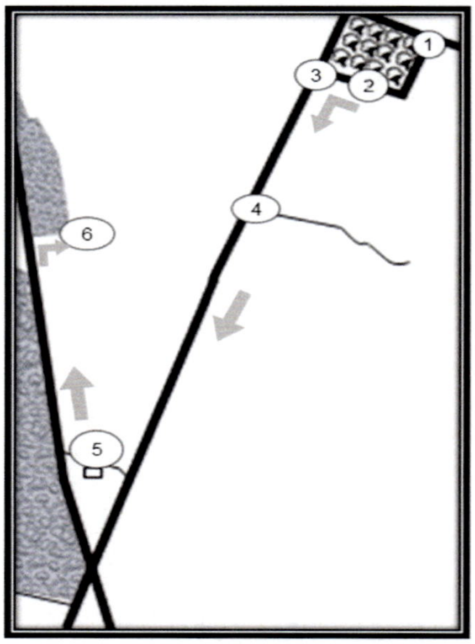

Quick Summary: You can see here how the Peach Orchard is prominent high ground compared to the open, rolling fields that drop off in each direction, especially behind you. The orchard rows provide space for men and wagons yet the trees in 1863 were high enough to offer concealment. A natural feature like this being located at the intersection of two critical roads enhances its importance.

The Union line was hinged on this point on July 2nd. It extended north to your right a mile along the Emmitsburg Road and southeast to your left beyond the woods to the Devil's Den (in the direction of Big Round Top).

The wayside sign in front of you features a painting by a correspondent who witnessed the battle plus the image of Confederate Brig. Gen. Joseph Kershaw, whose brigade made the first infantry attack against the Peach Orchard.

More Detail: Consult the narrative under Walking Tour Stop #1 on page 2.

Hiking Tour Stop #2: "Twin Threes" 3rd Mich./3rd Maine Flank Markers, South Side

- **GO:** Turn left and walk south along the eastern edge of the Peach Orchard to the southeast corner to the 3rd Michigan monument. Turn right and continue 30 yards to the pair of flank markers to give you a better vantage over the landscape.
- **ON THIS LEG OF THE WALK:** You are walking with the advance of two artillery units that moved to position on the edge of the orchard that you are approaching. You are also walking along with the charge of the 2nd New Hampshire Infantry through the orchard to your right.
- **STOP:** at the flank markers for the "Twin Threes" – the 3rd Michigan and the 3rd Maine.
- **LOOK:** south along the Emmitsburg Road, facing the Rose Farm.

Quick Summary: You have a commanding view of the Rose Farm in front of you and Warfield Ridge along the horizon to your front and right. This is the landscape across which the Confederate attack would begin on July 2nd.

The large square monument to your left, facing south, commemorates the 3rd Michigan Infantry regiment. It marks the end of the regiment's skirmish line, which continued down to the woods marking the Stony Hill.

As the artillery fight developed, that was also the position occupied by six guns of 1st New Jersey Battery B to your left and then by four guns of the 15th New York Battery at your location as they exchanged fire with Confederate guns along Warfield Ridge.

More Detail: Consult the narrative under Walking Tour Stop #2 on page 5.

Hiking Tour Stop #3:
2nd New Hampshire Corner/Emmitsburg Road

- **GO:** Turn right and walk west for 65 yards along the southern edge of the Peach Orchard.
- **ON THIS LEG OF THE HIKE:** You are passing along the combat line of the 2nd New Hampshire Infantry, which charged to this position to repulse the attack of two South Carolina units coming up from your left.
- **STOP:** at the painted post-and-board fence by the southwest corner where Birney Avenue intersects the Emmitsburg Road.
- **LOOK:** south along the Emmitsburg Road, facing the gap in the Warfield Ridge.

Quick Summary: The men of Kershaw's South Carolina Brigade began their advance from Warfield Ridge to your right, crossing the Emmitsburg Road toward the Rose Farm. Once across the road, a detachment of 800 men – two regiments plus one battalion – broke off and turned toward the Peach Orchard and Wheatfield Road.

The South Carolina attack moved against the position where you are now standing and against the south edge of the orchard along your left. The 2nd New Hampshire and 3rd Maine regiments came up from behind you, formed ranks on the south side of the Peach Orchard and repulsed part of the attack, driving the Confederates down the slopes and back up to the Rose Farm lane.

The thin obelisk marker to your right indicates this was the advanced position of the 68th Pennsylvania regiment, which held the line along the edge of the orchard that faces west.

More Detail: Consult the narrative under Walking Tour Stop #3 on page 7.

Hiking Tour Stop #4:
Rose Farm Lane

- **GO:** Walk south on Emmitsburg Road 300 yards, using the left shoulder of the road for safety to best see oncoming traffic. (Sometimes there is a mowed path inside of the fence. The problem is this path dead-ends in the creek bed.)
- **ON THIS LEG OF THE HIKE:** You are walking along the line of the withdrawal of the South Carolina units that hit the Peach Orchard and were driven back by the New Hampshire men. You are walking down into and then up back out of the brush-filled ditch in which form the headwaters of the West Branch Plum Run.
- **STOP:** at the dirt road that represents the Rose Farm lane.
- **LOOK:** to the right and face westward across the Emmitsburg Road.

Rose Farm

Quick Summary: Kershaw's Brigade crossed the Emmitsburg Road on this ground covering hundreds of yards to your left and right with four regiments and one battalion. A sixth unit, the detached 15th South Carolina regiment, crossed the Emmitsburg Road far to your left (south), separate from the rest of the brigade and destined to fight independently.

Once across the road, Kershaw would further divide his brigade. Three of his units totaling about 800 men wheeled to your right to face north. They launched their own, independent attack directly against the south side of the Peach Orchard and also against the Wheatfield Road to the right of the orchard. The attack failed. Kershaw led his two remaining regiments, the 3rd and 7th South Carolina, straight east across the Rose Farm behind you and against Union troops in the woods beyond it.

More Detail: There is a unique granite marker along the fence line across the street. It commemorates Kershaw's Brigade, indicating they "crossed the Emmitsburg Road in this area." It is quite different from the standard iron and bronze War Department tablets from the turn of the century. This one was erected in 1970 by "Project Southland," a private group of advocates for South Carolina history.

Kershaw's Brigade was not really centered on that monument. The center point of the brigade crossed the road somewhere to your left on the south side of the Rose Farm lane. Kershaw said First Corps Commander Lt. Gen. James Longstreet accompanied him as his brigade crossed the fields in front of you and came up to the road. Longstreet asserted personal control over the initial movement of the brigade.

Kershaw

Turn around and face eastward toward the Rose Farm.

From your left, the alignment of the brigade was as follows: 8th South Carolina regiment on the left flank, the 3rd South Carolina battalion next to them and the 2nd South Carolina regiment – the guide regiment – crossing a bit farther south of the Rose Farm lane to your right. To their right was the 3rd South Carolina regiment and the 7th South Carolina regiment on the right flank of this line.

Make a half-turn to your right to face south toward the white farmhouse 600 yards in front of you.

Kershaw had posted Col. William de Saussure's 15th South Carolina regiment in a detached position near the white Phillip Snyder farmhouse. They held the right flank of the line of artillery batteries firing in support of Major Gen. Lafayette McLaws' Division. When Kershaw's Brigade advanced, the detached 15th South Carolina crossed the Emmitsburg Road far to the south, separated from the rest of the brigade. The confusion was aggravated by the fact de Saussure would be killed early in the advance.

Why is "high ground" such a big deal?

In all eras of military history, from the conquests of Alexander the Great to the war in Iraq, the high ground represents a stronger position. Commanders on high ground can get a complete view of their enemy while concealing their own forces. Troops on high ground have superior "fields of fire" and present a smaller target than troops on low, open ground. Weapons of the 19th Century favored higher ground because they were more accurate when fired level. Lastly, an enemy that attacks must do so uphill.

Turn around and face back up the Emmitsburg Road, to the north, toward the Peach Orchard.

Once his men crossed the Emmitsburg Road over the ground you are on, Kershaw would divide his brigade. The 8th regiment, 3rd battalion and 2nd regiment wheeled toward the north to launch their own independent attack directly against the Peach Orchard and Wheatfield Road. The 2nd regiment's commander, Col. John D. Kennedy, was placed in overall command of this wing. He was a trusted deputy. Kershaw's first command was as the colonel of the 2nd South Carolina and Kennedy had been his lieutenant colonel. The ranks of this storied regiment included Sgt. Richard Kirkland, famous as "the Angel of Marye's Heights" where by some accounts he may have brought water to wounded enemy soldiers at the Battle of Fredericksburg, December 1862.

The ground in front of you witnessed the charge, bloody repulse and withdrawal of the 8th South Carolina infantry with part or all of the smaller 3rd South Carolina battalion. The regiment's attack began where you are, with the 3rd battalion on their right flank, but the withdrawal ended in the ravine below you. Eventually the regiment would move to their right to join in the fighting in the woods behind the Rose Farm. The second phase of the attack with the 2nd South Carolina infantry regiment would move to the right of the orchard and directly against the Union guns on the Wheatfield Road.

Turn right and face east again toward the Rose Farm.

Kershaw led his other two regiments, the 3rd and 7th South Carolina, on the straight line eastward through the Rose Farm and into the Stony Hill woods beyond it.

Historian James Hessler, esteemed author of "Sickles At Gettysburg" (Savas Beatie 2009) and a fellow licensed guide from the authors' same Class of 2002, makes a fascinating observation here. You can see that the Peach Orchard is the dominant high ground to your left. But in front of you, Little Round Top recedes from view, especially in July when the trees are at their fullest. Given these two vantages, Hessler suggests Major Gen. Daniel Sickles may have understood the Peach Orchard on this main thoroughfare to be more important high ground than the rocky hill off in the wilderness a half mile to his east.

Hiking Tour Stop #5: Snyder Farm

- **GO:** Continue walking south on Emmitsburg Road 600 yards until the snake rail fence on the right side of the road ends and a short section of white picket fence begins. Turn right and carefully cross the Emmitsburg Road. Go through the gap between the snake-rail and picket fence. Proceed to the dirt-path horse trail that continues west toward South Confederate Avenue.
- **ON THIS LEG OF THE HIKE:** You are walking across the front of Kershaw's Brigade as they crossed the Emmitsburg Road moving right to left. You are also walking across the line that Brig. Gen. George Anderson's Brigade crossed earlier and that Brig. Gen. Paul Semmes' Brigade would cross later.
- **STOP:** in the dirt path halfway to South Confederate Avenue.
- **LOOK:** back north across the fields toward the Peach Orchard 1,000 yards up the Emmitsburg Road.

Quick Summary: You are at the longest range for the Confederate guns of Col. Henry Coalter Cabell's battalion firing into the Union position in the Peach Orchard. The four rifled cannons of Capt. John Fraser's Pulaski (Ga.) Artillery would have been arrayed in front of you.

Looking to the right, you have a striking view of the Devil's Den and Little Round Top behind it, which from this vantage create the illusion of being one single mass of rising stone.

Your position also offers you the view of what would soon be the field of battle as first seen by Union officers riding in from Maryland on July 1st, including one Major Gen. Daniel Sickles. While you might expect a simple glance right would to have revealed Little Round Top as a citadel, eyes front on the Emmitsburg Road would cause the Peach Orchard to make a big first impression as the most valuable high ground.

More Detail: There were four batteries here in Cabell's battalion, which was attached to McLaws' Division. The four rifled cannons of the Pulaski Artillery were the far right end of Cabell's battalion. The other three batteries were in position across the open ground between you and the Tower Woods 500 yards to your front left.

Cabell

You are also in the vicinity of Col. de Saussure's 15th South Carolina regiment, probably located across South Confederate Avenue in the Biesecker Woods. Behind you in the Snyder Farm is the left flank of the Texas Brigade in Major Gen. John Bell Hood's Division.

Turn right to face east.

Thanks to restoration work by the National Park Service, you now have a striking and dramatic view of the Devil's Den and the rocky front face of Little Round Top behind it. (When the authors began their study of the battlefield in the 1990s, this view was obscured by forest.) If you use your imagination to remove from view the vehicles, people or monuments that may happen to appear on Devil's Den, you can see how for the Confederate troops in 1863 it was not immediately evident that Devil's Den and Little Round Top were two separate features. From this vantage they appear to be one single mass of stone.

This vantage also helps explain why Fraser's battery was the hardest hit of the four batteries on Cabell's line. On the left edge of the Devil's Den would be the Houck's Ridge position of Capt. James Smith's 4th New York Artillery. Firing from the front right corner of the Peach Orchard would be the six Parrott rifles of Capt. Judson Clark's 1st New Jersey Battery B, who in his report identified the Confederate battery near the white farmhouse as his first target. Thus Fraser's men would catch hell from two directions. The result was seven killed and the captain himself mortally wounded. The battery's losses in men and horses would be so heavy, their surviving guns would be folded into Capt. Basil Manly's North Carolina Battery A.

Your position also offers you the view of the Gettysburg battlefield first presented to Union officers using the Emmitsburg Road on their ride north from

Maryland, including Sickles himself. It would seem that a simple glance to the right here would have revealed Little Round Top to be a crucial defensive position. Yet when Sickles arrived, the battle he would have heard would have been occurring less than two miles in front of him. With eyes front toward the sound of the fighting, the Peach Orchard would have been unmistakable as the critical high point a half-mile ahead directly on the Emmitsburg Road. And as we have seen at Hiking Tour Stop #4, the farther you ride up this road toward the Peach Orchard, the more Little Round Top recedes from view.

Hiking Tour Stop #6: South End Tower Woods

- **GO:** Continue on the dirt path west a short way to South Confederate Avenue, turn right and go north on the asphalt road 500 yards until you reach the Tower Woods on the right side of the stone wall. Turn right, step carefully over the stone wall and walk east along the edge of the Tower Woods. (A path might be mowed along the field side of the wall if you prefer.)
- **ON THIS LEG OF THE HIKE:** You are walking along the back of Cabell's whole battalion line, an area that would have been crowded with limbers and caissons – and harassed by overshots from the Union batteries in the Peach Orchard. The stone wall you are following was there, as were the Biesecker Woods across the street, but the road itself did not exist.
- **STOP:** at the front corner of the Tower Woods at the point where the Peach Orchard comes into view.
- **LOOK:** across the fields toward the Peach Orchard.

Quick Summary: You have come to the position of Cabell's batteries closest to the Peach Orchard some 600 yards northeast of you. Here either Cabell or his artillery commander, Col. E. Porter Alexander, deployed the battalion's shorter-range bronze guns.

The "gunfight" at the Peach Orchard grew in intensity. One veteran gunner, Anderson Reese of the Troup Artillery from Georgia, said, "The firing was the most rapid I ever witnessed. The ground literally vibrated under the continuous roar."

You are also at the southern edge of the Tower Woods in which half of Kershaw's Brigade was positioned. When the brigade advanced they moved across the field in front of you toward the Emmitsburg Road. Kershaw in his report said Lt. Gen James Longstreet accompanied him on the advance in order to see the ground beyond the Emmitsburg Road and to direct in person the movement of Kershaw's troops.

More Detail: Cabell's shorter-range bronze guns were 12 pounders, two of them Napoleons and two of them howitzers. In fact, they blended a section of Capt. Henry Carlton's Troup Artillery with a section of Capt. Manly's North Carolina battery to make sure the battalion's four bronze guns were positioned closest to the Peach Orchard. The 10 longer-range iron rifles of the battalion were positioned to the right for the longer shot.

McLaws ordered this deployment of his artillery battalion initially so the return fire from the Union batteries would not hit Kershaw's men. When the brigade advanced they moved across the field in your front toward the Emmitsburg Road, with the brigade roughly centered on the Rose Farm lane where you were standing for Hiking Tour Stop #4.

This area can also be seen as the extreme left flank of Hood's division. For sure, the Texas Brigade, which led the attack, was quite far to your right at the Snyder Farm and straddling the Emmitsburg Road. But supporting the Texans was Brig. Gen. George "Tige" Anderson's Georgia Brigade positioned in the Biesecker Woods behind you. When Anderson advanced his brigade, his left-most regiment was the 9th Georgia infantry. It passed close by your position, according to the detailed account provided by one officer, Capt. George Hillyer, who would end up in command because of the loss of so many

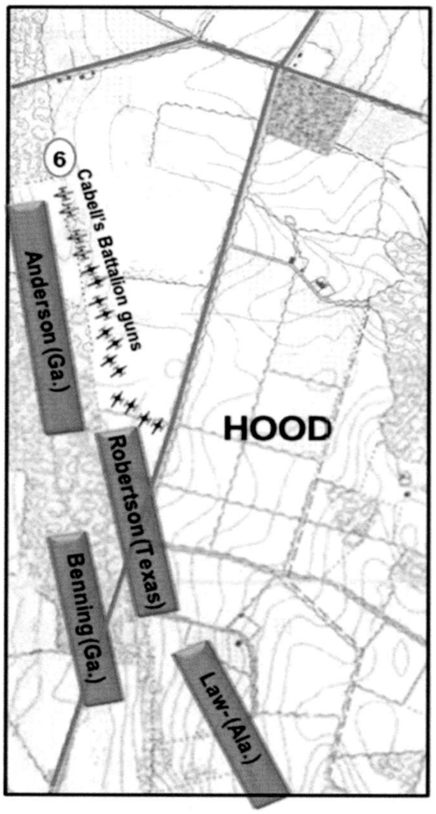

officers. Hillyer said his men marched directly through the guns manned by his fellow Georgians of the Troup Artillery.

Anderson's Brigade went in before Kershaw. Anderson followed the left flank of the Texas Brigade all the way to low ground in front of the Devil's Den before veering left to engage Union regiments in the Wheatfield. Kershaw, in turn, would come in on Anderson's left.

Hiking Tour Stop #7: Louisiana Memorial

- **GO:** Turn around and go back to South Confederate Avenue, climb back over the wall and turn right. Walk about 0.4 mile north past the War Dept. Observation Tower and many gun batteries. Carefully across the Millerstown Road.
- **ON THIS LEG OF THE HIKE**: You are passing behind the original battle line of Kershaw's Brigade and then as you pass the War Dept. tower the battle line of Barksdale's Mississippi Brigade. You are passing behind or through four batteries of Col. E. Porter Alexander's reserve battalion. You are crossing the spot that provided McLaws with his long-awaited first glance at the battlefield and the Peach Orchard. It would be an unhappy moment.
- **STOP:** at the Louisiana Memorial, which features the sculpture of a female figure blowing into an immense horn.
- **LOOK:** east across the fields toward the Peach Orchard and the Emmitsburg Road 550 yards in front of you.

Quick Summary: You are near the point on Warfield Ridge where Major Gen. Lafayette McLaws arrived and was shocked to discover the presence of Union brigades in the Peach Orchard in front of you and left along the Emmitsburg Road.

Your view of the Peach Orchard and the Sherfy Farm along the Emmitsburg Road is quite clear – but you can see nothing beyond it. You cannot see anything or anyone on the fields on the other side of the Emmitsburg Road. Neither could troops, gunners or their commanders, including Longstreet, McLaws and perhaps Gen. Robert E. Lee himself. This limitation adds to the crucial and dangerous nature of the Peach Orchard position.

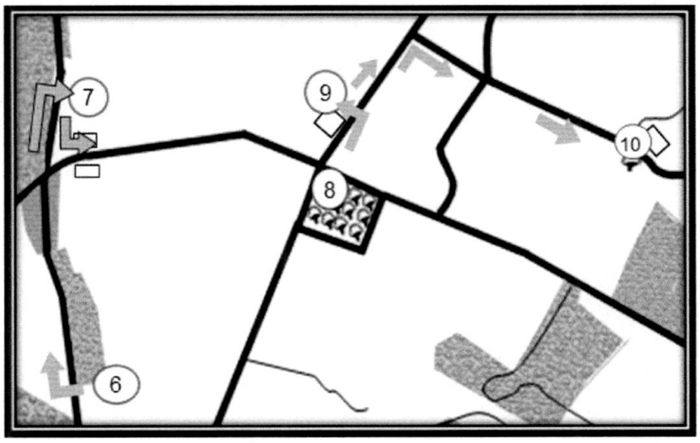

You are also standing on the ground where the opposing batteries were at their closest during the artillery "gunfight" at the Peach Orchard. The four howitzers of the Branch Artillery of South Carolina were at your location with four bigger howitzers of the Madison Artillery of Louisiana to your right near the Millerstown Road. Both were 550 yards away from the Union batteries in the Peach Orchard.

It is here that Alexander would say: "I hoped to make it short, sharp and decisive, for at close range there was less inequality in our guns and especially in our ammunition, and I thought if ever I could overwhelm and crush them, I would do it now."

This position was also the line from which Brig. Gen. William Barksdale would later lead his four Mississippi regiments forward to strike the Peach Orchard. The Mississippi Memorial stands 75 yards to your right.

More Detail: The four howitzers of the Branch Artillery of South Carolina, commanded by Lt. Stephen Gilbert, are less than 600 yards away from the Peach Orchard. The combination of the short range and the weight of metal deployed are astonishing.

Gilbert's and another battery of four 24-pounder howitzers near the Millerstown Road intersection, the Madison Artillery under Capt. George Moody, were both part of Alexander's battalion of reserve artillery. Two more of Alexander's batteries were on the south side of the Millerstown Road.

McLaws

Regarding the positioning of Barksdale's Brigade for his famous charge, you are located at the likely position of the 13th Mississippi regiment. This was Barksdale's original infantry command and so this may have been the spot at which the Mississippi firebrand rallied his men, mounted and shouting while waving his hat in the air. The 18th Mississippi was at the end of the line to your left, the 17th Mississippi was to your right and the 21st Mississippi, at the right flank of the brigade, was straddling the Millerstown Road all the way up to the Tower Woods.

Before all the fighting started, McLaws first came up to see the ground for himself. He had been leading his division across the fields and farm roads behind you for many hours, including a circuitous countermarch to avoid detection by Union observers. You are now in the vicinity where McLaws was standing when he discovered the presence of a substantial Union line about 600 yards to his front. It had been expected that some Union troops would be in the way but not a force of this size. (In war, surprise is seldom welcome.)

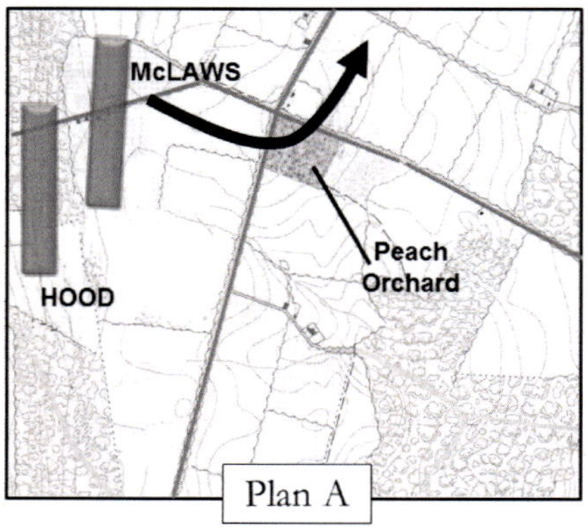

Plan A

Now McLaws sent an officer back to alert Longstreet to the emergency. The situation had changed. A message was returned to McLaws insisting there could be nothing more than a blue regiment in front of him and that McLaws should just brush it aside. McLaws had this messenger return to Longstreet with the insistence that the corps commander was failing to grasp the new conditions on the ground. This prompted Longstreet to come up in person and see the situation for himself.

Taking stock, Longstreet quickly applied the adage familiar to old warriors: "The first casualty of war is the plan."

Plan A was to have McLaws move first by forming his line at the Peach Orchard, facing north and using the Emmitsburg Road as a guide. He was to drive in the Union flank. Hood was to come up in support. Then the divisions of the Third Corps would engage from the Seminary Ridge woods to your left.

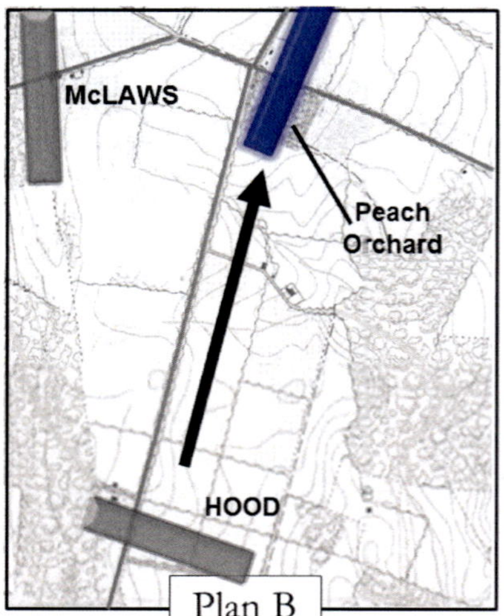

Plan B

Plan B switched McLaws with Hood. Hood would form his line a half-mile farther south along Warfield Ridge itself, near where you stopped for Hiking Tour Stop #5. Hood was to turn north and attack by marching up the Emmitsburg Road through the Rose Farm and the Peach Orchard. Hood's Division was to drive in this new Union flank. McLaws was to follow in support, with the Third Corps division of Major. Gen. Richard Anderson moving up on their left in support.

The movement of Hood's Division to a new position and the alignment of his four brigades into an attack formation would take time (See map on page 20). During this time, some 30 to 45 minutes, Confederate and Union artillery batteries started banging away at each other with ever more guns on both sides coming up, unlimbering and opening fire. Of his Union opponents Alexander would say: "They really surprised me, both with the number of guns they developed and how they stuck to them."

At the height of the artillery action, 48 Confederate guns along the length of Warfield Ridge were hammering away at 38 Union guns in position from the Sherfy Farm through the Peach Orchard to the Devil's Den. The Sherfy Farm is represented by the large red barn on the Emmitsburg Road in front of you with the two-toned Sherfy Farmhouse to the left of the barn. The firing distance to the Peach Orchard and other points on the Emmitsburg Road was 550 yards muzzle to muzzle. This action between 86 guns made it the largest close-range artillery "gunfight" of the entire Civil War.

In the midst of the sudden rearrangement from Plan A to Plan B, it is interesting to note the high level of activity of "Pete" Longstreet himself. First he moved up with Kershaw's Brigade to the Emmitsburg Road. Then he ventured out onto Barksdale's forward skirmish line marked off by the fences 250 yards in front of you. This was an unusually hazardous undertaking for a lieutenant general and corps commander.

Longstreet would be in line at your position with Barksdale as the Mississippians prepared to attack. Behind Barksdale's Brigade was a second Georgia brigade under the command of Brig. Gen. William Wofford. Longstreet would accompany Wofford's Brigade as they marched after Barksdale right through your present location, into the Peach Orchard and all the way down through the Wheatfield to the base of Little Round Top. Longstreet was there on the line with his men when he decided to call off the attack at the end of the day.

Hiking Tour Stop #8:
Peach Orchard/1st New York Battery G

- **GO:** Return to the Millerstown Road intersection 130 yards south, turn left onto the Millerstown Road, walk 550 yards to the Emmitsburg Road intersection. Use the broad grass shoulder on the right side of the road. Carefully cross the Emmitsburg Road into the Peach Orchard.
- **ON THIS LEG OF THE HIKE:** You are walking precisely with the line of attack of the 21st Mississippi infantry from Warfield Ridge aiming directly into the Peach Orchard. The point where the Millerstown Road bends the farthest left (north) marks the spot at which an old farm lane once led off toward the John Staub Farm, the scene of heavy skirmish fighting. That lane and the Staub Farm no longer exist.
- **STOP:** at the northwest corner of the Peach Orchard at the monument for 1st New York Battery G flanked by the two green Napoleon cannon.
- **LOOK:** across the Emmitsburg Road toward the west the way you just came.

<u>Quick Summary:</u> Your walk has brought you through the field of brisk artillery fire to the position of 1st New York Battery G, which traded shots with the stronger line of Confederate batteries at the positions along Warfield Ridge where you made your last three stops.

You are also facing the ground across which Barksdale would lead the powerful charge of his brigade. While Kershaw's men failed to break the Union hold on the Peach Orchard, Barksdale's Brigade would smash it. By the time that attack came, the New Yorkers' Napoleon guns had been replaced by the rifles of 1st Pennsylvania Battery C & F. (That battery's twin monuments are behind you to your right.)

The 21st Mississippi battle line would march straight into the Peach Orchard over the point where you are standing. The 2nd New Hampshire regiment would change position to face this attack from the west.

<u>More Detail:</u> Consult the narrative under Walking Tour Stop #4 on page 9.

Hiking Tour Stop #9: Sherfy Farm

- **GO:** Carefully re-cross the Emmitsburg Road intersection, turn to the right and walk north up the left shoulder of the Emmitsburg Road about 250 yards to the Sherfy Farm, then turn left away from the road and into the farm, passing in between the barn on your left and the farmhouse on your right.
- **ON THIS LEG OF THE HIKE:** You are walking along the line of two Pennsylvania regiments, the 68th and the colorful 114th, "Collis' Zouaves," who were position to the left and within the grounds of the Sherfy Farm complex. These are the units that would be smashed and turned out by three of Barksdale's Mississippi regiments.
- **STOP:** at the back of the Sherfy House where a picnic table or chairs might be available. (On most days the house is unoccupied so feel free to take a load off.)
- **LOOK:** west across the fields toward Warfield Ridge and the Louisiana Memorial.

Quick Summary: You are at the front line of the battle-hardened veterans of the 114th Pennsylvania, "Collis' Zouaves," where they stopped the attack of the 18th Mississippi infantry directly in front of them. But the Union regiment to their left, the 68th Pennsylvania, gave way under the impact of two larger Mississippi regiments. As the Peach Orchard position collapsed, Barksdale turned these two regiments toward the north – directly onto the flank and rear of the 114th Pennsylvania.

The Zouaves' line was routed. The 57th Pennsylvania fled from the Sherfy House and yard area while the 105th Pennsylvania, farthest from the flanking attack, withdrew to reform on the Trostle Farm lane, which is present-day United States Avenue, the next avenue to the north intersecting the Emmitsburg Road.

Cavada

More Detail: Collis' Zouaves were formed and commanded by Col. Charles Collis, who was ill at home during the Gettysburg Campaign. Lt. Col. Frederico Cavada had the command. A native Cubano, Cavada's American mother had returned to Philadelphia when her husband died. (Cavada's kid brother Adolph was just up the Emmitsburg Road on Brig. Gen. Andrew Humphreys' staff.) The Zouaves were a fierce, respected fighting unit that had plenty of good cover on the fences, buildings and sheds of the Sherfy Farm.

Some 150 yards in front of them, Mississippi skirmishers crawled along the fences and in the recesses of Staub Farm lane, including one James Duke, who had an older brother, Archibald, in the same regiment. James peered across the deadly ground from behind a fence post when he sensed the presence of a heavily bearded officer near his side. He glanced up to encounter none other than Gen. Longstreet himself. Duke would write that he told the general there were three lines of infantry, followed by the question: "General, do you think we can take those heights?"

He said Longstreet's reply was this: "I don't know. Do you?"

Behind them, the Mississippi Brigade endured what unit historian John McNeily described as "the severest of all tests on troops, to receive (artillery) fire without returning it." When these regiments were finally unleashed, they shattered the Peach Orchard position. An officer in a New York regiment nearby said they started receiving fire in their rear from the Mississippi rifles. Adolph Cavada said of Frederico's unit, "The enemy's columns were upon them before they

Sherfy house with 1863 outbuildings

could fall back. All was confusion on that side."

The 57th Pennsylvania was using the farmhouse behind you as a fortress, shooting from all the windows. Their rate of fire was so loud, Capt. Alanson Nelson had to go from man to man to yell to each one that it was time to get out. By the time Nelson himself got out, Mississippi men had overrun the yard. Faced with surrender, Nelson chose to sprint, hurtling a fence amid a volley of fire. "Either their aim was poor or else I outran their shots, for they never touched me. As each shot would whiz past me, I increased my efforts," Nelson said.

The Zouaves suffered 57 killed and wounded with about the same number taken prisoner, of whom 44 would perish while being held in prison camps such as Andersonville. Cavada was made a prisoner and on his release from Libby Prison in Richmond, Va., he was brought up on charges of cowardice and removed from the army.

About a dozen of the wounded Zouaves would be brought into the Sherfy Barn as a temporary hospital. They perished when the barn caught fire and burned to the ground. Commenting on the smoldering remains, one New York soldier wrote that it "was a scene more than unusually hideous... The crisped and blackened limbs, heads and other portions of bodies lying half consumed among the heaps of ruins and ashes made up one of the most ghastly pictures ever witnessed, even on the field of war." Some of their graves are in the Pennsylvania section of the National Cemetery under the engraving: "Unknown Zouave."

On the Confederate side, James Duke would search past dark until he located Archie wounded in a hospital. Archie was so pleased to see that his little brother was safe, he showed little concern for his own badly wounded arm. The arm became gangrenous and the amputation killed him.

ORIENTATION: You have reached the halfway point of your four-mile hiking tour. If you have had enough for today, re-cross the Emmitsburg Road, walk south on the left shoulder back to the Peach Orchard, turn left on Wheatfield Road and return to your car.

Hiking Tour Part II – Stop #10: Trostle Farm

- **GO:** Return to the Emmitsburg Road, turn left and walk north up the left shoulder of the Emmitsburg Road to the intersection with United States Avenue. Turn right, carefully crossing the Emmitsburg Road and walking east along the right shoulder of United States Avenue about 0.25 mile past the Sickles Avenue intersection and down hill to the Trostle Farm.
- **ON THIS LEG OF THE HIKE:** You will walk past the monuments of the 114th Pennsylvania and 57th Pennsylvania regiments in the front yard of the Sherfy Farm and then the 105th Pennsylvania "Wild Cat" regiment monument at the United States Avenue intersection where you will cross the Emmitsburg Road. Then you will walk the final line of the Wild Cat Regiment as it held off Barksdale's charge, backed up by the 11th New Jersey, the 120th New York and other Excelsior Brigade regiments who formed their line behind the Klingel Farm to your left.
- **STOP:** at the pair of green Napoleon cannon in front of the Trostle farmhouse.
- **LOOK:** southwest across the road and fields to the close horizon.

Trostle Barn with dead horses 1863

Quick Summary: You are on the ground of the climactic last stand of Capt. John Bigelow's 9th Massachusetts Battery. The Napoleon guns represent the battery's second position after withdrawing across the fields in front of you. Bigelow's battery was then ordered to sacrifice itself in order to stunt the advance of the Mississippi infantry.

Cross the street and look up the slope.

You can see that the view of the horizon is quite limited at this spot. You are at the bottom of a half bowl. This short horizon enabled Bigelow's gunners to keep the Confederates at bay by firing canister, solid cannonballs and case shot with short fuses. Despite these frantic efforts, the battery was eventually overrun with most of their officers and a third of the men killed or wounded.

9th Massachusetts Battery monument in their first position on Wheatfield Road

More Detail: Understanding the modern and historic lane here can be tricky. The remnant section of stone wall with the 9th Massachusetts Battery marker, a granite depiction of a limber chest, appears on the left side of the modern park road. However, it marks the right edge of the historic lane that has long since vanished. The 1863 Trostle Lane ran between the stone wall and the fence in front of the house. The small marker braced by the pair of Napoleon cannon represents the position of "Bigelow's Gate," a space in the wall through which his guns tried but failed to escape as the battery was overrun.

Ordering Bigelow's battery to sacrifice itself would seem an act of madness – but hold on. Advancing troops also could not see you until they cleared the top edge of the bowl less than 150 yards away – just in time to meet lethal sheets of iron. Bigelow's gunners fired canister, they fired case shot with short fuses and they skip-fired solid rounds across the ground to keep the attacking infantry at bay.

This momentary geographic advantage must not be overstated. Untested in battle before July 2, the gun crews of Bigelow's battery exhibited exceptional ability, courage and devotion to duty, especially the junior officers, all of whom were killed or wounded. Bigelow himself was badly wounded in the gut and had to be led off the field on his horse by his bugler, who received the Medal of Honor for his valor.

Aided by Alabama men firing and charging down the Trostle Lane to your right, the Mississippi men overran Bigelow's guns, capturing five of them – but, alas, only for a short while. The crew of one gun managed to limber up their piece and to get out by banging over a small stone wall that can still be found to your left hidden by the tall grass and brush.

Hiking Tour Stop #11: Plum Run Line

- **GO:** Continue east on United States Avenue about 100 yards to the intersection with the dirt John Weikert Lane (today a bridle path) bordered by snake rail fences. Turn onto the dirt lane and follow it across the Plum Run foot bridge and then up the slope another 90 yards onto the low ridge by a cluster of trees.
- **ON THIS LEG OF THE HIKE:** You are walking through the ridge on which Col. Freeman McGilvery formed his "Plum Run Line" of guns.
- **STOP:** at the cluster of trees on the low ridge.
- **LOOK:** back in the direction you came toward the Trostle Farm.

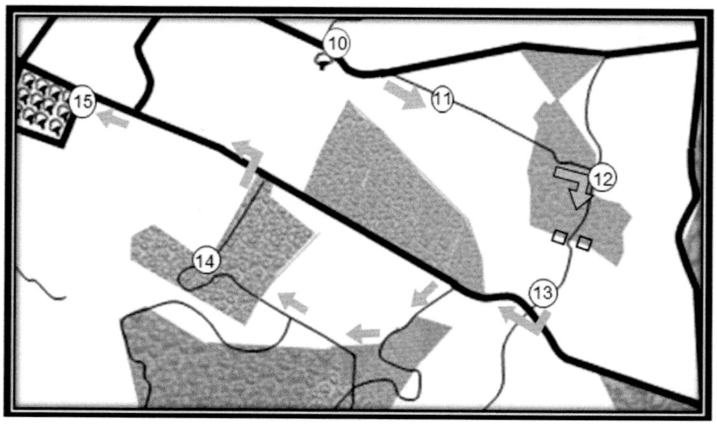

Quick Summary: You are on part of the ridge on which Lt. Col. Freeman McGilvery formed his so-called "Plum Run Line" of guns to stop the Confederate attack once and for all. It includes the position of 5th United States Battery I, which probably straddled the farm lane you are on (with the fences removed.) This battery of U.S. Army regulars, commanded by a newly minted West Point graduate, held the left end of McGilvery's line.

McGilvery's line essentially formed to the right (north) from your position across United States Avenue (which did not exist) and into the fields beyond. This might be something of an oversimplification. One report from a battery commander, Lt. Edwin Dow, said his guns were positioned closer to the woods behind you to the right. Also, it has never been entirely clear which batteries or remnants were in this group since McGilvery picked off sections of guns as they were retreating past the Trostle Farm, ordering them into position despite the fact that some of those guns were not under his command.

Whatever the truth about such details, the fact was that a concentration of iron and bronze muzzles formed here in support of one another and they were all aimed in the same direction – straight at the Mississippians by the Trostle Farm and on the adjacent field. And this whole gun line was under the command of a single man, McGilvery, a 39-year-old sea captain from Maine who knew how to handle guns, how to handle men and how to keep his cool under immense pressure.

McGilvery

Barksdale

More Detail: As McGilvery's Plum Run Line let loose with a sheet of flame, the Mississippi attack that started a mile back on Warfield Ridge would be stopped but cold. It was the farthest point of the Confederate advance in this section of the battlefield on July 2nd.

McGilvery's guns were deadly in their own right but they were also something of a soldier's trick. It was customary military procedure to always assign infantry to protect artillery. If guns were in evidence on a battlefield, you had to expect there were troops nearby in support. If the Union had a line of guns from five or six batteries across that ridge, there had to be a whole brigade of Union infantry in the woods right behind them. There was of course no one behind them at all. The woods were for the moment empty. McGilvery's intuition, skill and willingness to gamble with a line of cannon were the only barrier blocking a disastrous Confederate breakthrough to Cemetery Ridge and across it to the Taneytown Road. (If Lee's troops could have gotten through here, they might have split the Union Army of the Potomac in half.)

There was no end to the drama. The guns positioned in the farm lane, 5th U.S. Battery I, would be overrun. Some of the men from the 21st Mississippi would get through the muddy thicket and across Plum Run and were able to charge up this slope to capture those guns. They might have been helped by the fire or bayonets of some of Kershaw's South Carolina troops who were coming through the Trostle Woods across the marshy ground to your left. The U.S. battery's young officer, Lt. Malbone Watson, would be badly wounded in the leg, which he would lose.

Turn to look to your right (north).

In the fields and thickets across United States Avenue and behind the Trostle Farm, an entirely separate and much larger fight would unfold. The three original regiments of Barksdale's Mississippi Brigade would shatter the Union line behind the Klingel Farm, then pivot once again to face toward the east in pursuit of retreating Union troops. This was the third and final leg of Barksdale's famous charge as his men formed battle lines and advanced from the Klingel Farm on the Emmitsburg Road eastward toward Cemetery Ridge, which is about 500 yards to your right (east). They had a clear view of the Michael Frey Farm on the Taneytown Road. If they could reach it, they would split the Union army's defensive line asunder.

Willard

"The Harper's Ferry Cowards" stood in their

way. Freshly redeployed, this brigade of four proud New York regiments – many of whom had been made prisoners in the humiliating surrender of Harper's Ferry the previous September – now had something to prove. Three regiments under Col. George Willard advanced with fresh vigor and full loads of ammunition to confront Barksdale's men, who for all of their valiant success had to be running short of breath and bullets. The two lines collided with ferocity on the open ground behind the Trostle Farm leading down into the thickets across from you. As on the Plum Run Line, the Mississippi attack over there would be stopped cold. The vainglorious Barksdale was shot several times, fell from his horse and was taken prisoner, mortally wounded.

Turn right to face east toward the woods. United States Avenue should now be on your left.

Willard attacked with three regiments. He dispatched his fourth, smallest regiment, the 39th New York infantry, to a separate task. The immigrant warriors of the so-called "Garibaldi Guards" charged from the woods in front of you and at the point of the bayonet they recaptured the Union guns where you are standing. Their charge was so vigorous, they trampled one of their own officers, Capt. Carlos de la Mesa, after he had fallen with a wound in the foot.

It should be noted that conventional wisdom for generations has held that McGilvery's July 2nd artillery line was formed not on this ridge at all but farther north and closer to Cemetery Ridge itself. That position is marked today by a U.S. government monument to 5th U.S. Battery I closer to present-day Hancock Avenue.

A growing number of experts, most notably NPS historian Eric Campbell, contend McGilvery's line was not by the monument but was positioned on the ridge where you are located.

Hiking Tour Stop #12: Weikert Woods

- **GO:** Continue to follow the dirt John Weikert Farm lane (and modern horse trail) east about 200 yards until you enter the woods. Continue on the trail another 160 yards until you reach the spot where the trail comes back out of the woods and bends sharply left.
- **ON THIS LEG OF THE HIKE:** As you enter the woods, the men of the 39th New York infantry would have been rushing forward through the trees at your left. Those immigrant warriors of the so-called "Garibaldi Guards" used the bayonet to recapture the Union guns of 5th U.S. Battery I.
- **STOP:** where the dirt trail bends sharply left. In 1863 this was a "T" intersection with another farm lane leading off to the right as well. This 1863 lane connected the George Weikert Farm to your left with the John Weikert Farm to your right.
- **LOOK:** forward in the direction you have been walking, toward the east.

Quick Summary: You are now standing in the vicinity of where Sickles parked some of his regiments hours before any fighting occurred on July 2nd. You can see this is low ground with poor fields of fire.

Sickles pitched his headquarters tent north of here along Cemetery Ridge late on July 1 but his mind was fixed on the Emmitsburg Road, on which he had just arrived, and on the Peach Orchard, which he had just crossed. He never grasped his part of the "fishhook" defensive line Major Gen. George Gordon Meade had envisioned for July 2nd. The confusing terrain did not make such a position evident.

Sickles

Look right in the direction of Little Round Top. You cannot really see it – which also means anyone up there cannot see you. This helps explain how the Confederate scout who made the early morning reconnaissance there reported back to Gen. Robert E. Lee that the position was open and undefended.

More Detail: This is crucial ground for understanding the advent of the fighting on July 2nd. Conventional wisdom for the Battle of Gettysburg – that veritable "library" of books and maps about the battle – make repeated and constant reference to the "fishhook" position from Culp's Hill to Cemetery Hill and then down the line of Cemetery Ridge to Little Round Top.

A reconsideration shows that on July 2nd the fishhook position existed in only one place: the imagination of the Union army commander. Meade planned to have a battle line from Cemetery Hill all the way to Little Round Top. It was not formed in reality.

The Union position on July 2nd was in truth more of a "horseshoe" from East Cemetery Hill and bending around to Cemetery Ridge – but no farther south than the present-day site of the Pennsylvania Memorial, where an entire Second Corps division was placed in a column, not a defensive line. Meade may have conceived of the idea that Sickles would form his men in a neat north-to-south line as a continuation of the Cemetery Ridge position to be occupied by Second Corps. But it did not happen on July 2nd.

The ground around your present location helps explain why.

Look around. There is no "ridge" of any kind in evidence. There is no way an untrained officer can conceive of a defensive line extending south from Cemetery Hill, which cannot be seen from here. Sickles would set up his first headquarters in view of the position assigned to Second Corps but his mind was fixed on the Emmitsburg Road and the Peach Orchard. He never grasped the "fishhook" position that Meade had envisioned.

Sickles had six brigades in his two divisions. He had split them into three units of two brigades each. One section of two brigades had spent the night back in Emmitsburg. Two other brigades under Union Brig. Gen. Andrew Humphreys went into camp last night in the dark on the sloping fields north of United States Avenue. On the night of July 1st, Humphreys' men had marched in the dark the wrong way. They arrived on the battlefield in a confusing countermarch from the west via the Millerstown Road, crossing the Emmitsburg Road and going into camp in the open fields east of it. They were in no way connected to Cemetery Ridge or to anything conceivable as a battle line.

Sickles' other section of two brigades under Major Gen. David Birney were the first to arrive. They camped in regimental groups by the George Weikert Farm to your far left and on the ground around your present location, forming no real line whatsoever.

Look to your right.

You cannot see much and you cannot see Little Round Top at all. The woods and Munshower Hill block your view. If you were to take the time to walk around and explore, you will see that this remains true from many vantages. At some points you may spot the woods that form the northern side of Little Round Top but at no point will you be able to see the open plateau of flat boulders that form the top of the hill and which reveal its importance as a defensive citadel. Add to this

Little Round Top 1883 View

the fact that by several accounts, July 2nd began with a low, soupy fog that persisted for several hours.

This also explains how the Confederate scout, Capt. Samuel Johnston, could make his infamous early morning reconnaissance of Little Round Top and report back to Gen. Lee that the position was open – setting off 150 years of controversy and debate for everyone. (Everyone, that is, but Capt. Johnston himself, who in post-war writings denied any supposed confusion.)

This is another point of conventional wisdom expressed in countless narratives about the Battle of Gettysburg: That Johnston could not have gotten to Little Round Top. How could he have failed to see the Union forces at the base of the hill? How could he have failed to see the imposing fishhook line? This conventional wisdom holds that Johnston must have been deluded, dishonest or must have gone up the wrong hill and then proceeded to file a gravely erroneous report with Lee that cost him the war.

It has always been ridiculous to imply that such a superb engineer and tactician as Robert E. Lee, legendary for his scouting adventure at Cerro Gordo, would have in his employ such an incompetent subordinate.

The fact that you cannot see the open position of Little Round Top from anywhere around here also means that anyone standing at the open position on Little Round Top cannot see you. The Twelfth Corps units initially assigned to this area late on July 1st had left the ground at daybreak, an hour or more before Johnston made his reconnaissance. If Johnston had been able to see any of Sickles' men in the vicinity of your current position, they certainly were not close.

Johnston never reported to Lee that he had failed to see any Union troops anywhere in Pennsylvania. He had slipped past deadly skirmish lines and Union cavalry patrols to conduct his reconnaissance. Of course Johnston saw hostile blue bellies, plenty of them. But his report was accurate that Little Round Top and the ground around it was open and unoccupied at 7 a.m. The hilltop would remain empty of combat troops for most of the day.

Hiking Tour Stop #13: John Weikert Lane

- **GO:** Leave the horse trail and turn right, heading south toward the out-building and yellow John Weikert farmhouse that you can see through the trees. The remnant farm lane here has wheel ruts but is overgrown. It is usually mowed but not always. This can be the roughest part of the Hiking Tour but it is short. Be cautious of abandoned cattle grates still in the ground. Pass out of the woods, walk between the farmhouse on the left and barn on the right. (This home is rented to a private family so pass through quickly.) Continue south to the gravel farm lane.
- **ON THIS LEG OF THE HIKE:** The rough path you will use was an open farm lane in 1863 linking the George Weikert Farm behind you to the John Weikert Farm in front of you. This lane was used by three brigades of Brig. Gen. John Caldwell's division coming from Cemetery Ridge to recapture the Wheatfield.
- **STOP:** where the John Weikert Farm gravel lane intersects the asphalt Wheatfield Road.
- **LOOK:** forward in the direction you have been walking, toward the south and the Crawford Avenue park road.

Quick Summary: You are now facing the Plum Run valley between Houck's Ridge to your right and Little Round Top to your left. This valley has come to be known as the "Valley of Death," although the origin of this nickname is a point of conjecture.

As you come up the lane from the John Weikert Farm buildings you could see for the first time the open plateau of Little Round Top. Wherever Sickles may have placed the first two brigades of Birney's division, they certainly were not here. So again, there was nothing for the Confederate scout Johnston to see.

The open, soggy ground to your right rear extending from the Weikert Lane all the way out to the Trostle Farm should really be considered part of the Wheatfield battlefield. Second Corps units formed their initial battle lines on this ground to advance across Plum Run and across the Wheatfield Road against the Confederate Georgia Brigade within the Wheatfield itself.

At day's end, the ground around you is also where the advance of the Georgia Brigade of Brig. Gen. William Wofford would come to a stop. Wofford's regiments had followed Barksdale's men into the Peach

Orchard but continued straight down the Wheatfield Road. Wofford's men attacked across the valley in front of you, right to left, against Pennsylvania Reserves on the base of the northwest side of Little Round Top. Once more, Longstreet was on hand to guide this brigade and he now ordered a halt to the fighting, over the objection of some officers, because he saw too many Union troops moving into position to your left and rear. These were more Pennsylvania Reserves and elements of the massive Sixth Corps just arriving on the battlefield.

Hiking Tour Stop #14: Stony Hill Loop

☛ **GO:** Turn right and walk west on the left side of Wheatfield Road. Use the left side of the road facing traffic. Pass over the Plum Run bridge and turn left onto Ayres Avenue to enter the Wheatfield. Usually trails are mowed to the 27th Connecticut monument (the thin obelisk with a bronze eagle on top.) Such trails are the best way to cross the Wheatfield. If there is no mowed trail, follow Ayres Avenue south 250 yards. When the asphalt road turns sharply left into the woods at the monument to the 148th Pennsylvania, leave the asphalt and walk to the right, cutting across the grass field to get to Sickles Avenue.

Once you are back on the asphalt, turn right onto Sickles Avenue and follow the road west into the woods. You will pass between the monuments to the Irish Brigade on the left and the 5th Michigan on the right. When Sickles Avenue bends sharply left, you can leave the asphalt road and remain straight onto footpath shortcuts that will bring you up the hill and back onto Sickles Avenue at the other end of "The Loop."

☛ **ON THIS LEG OF THE HIKE:** You are walking across the Wheatfield, the bloodiest ground in North America. As you re-enter the woods you are passing the most famous of the monuments to the Irish Brigade with its Celtic Cross and Irish wolfhound. You will follow their attack up the short, wooded slope to Stony Hill.

☛ **STOP:** across Sickles Avenue to the left of the 116th Pennsylvania monument.

☛ **LOOK:** west up the slopes and open fields toward the Peach Orchard.

Quick Summary: No other point on your hike makes it as clear as this spot that the Peach Orchard is the high ground. It is also clear that from down here that you cannot see beyond it. Troops, gunners and commanders could not see anything or anyone beyond that ridge of high ground, including any of the Confederate brigades arriving on Warfield Ridge and what they might be doing.

The woods to your left and front have come to be called the Stony Hill. This is where Kershaw led his two center regiments, the 3rd and 7th South Carolina, across the Rose Farm beyond the woods to your front left. In their first attack, Kershaw's men would drive Union troops through your location and back to your right.

The ground before you that slopes up toward the Peach Orchard was the scene of the disastrous change of direction taken by some of the South Carolina men who were trying to advance on the Union batteries holding the Wheatfield Road to your right. Confused orders caused them to change direction and march directly at you, in the process exposing their flank to the devastating short-range fire of 12 cannon.

More Detail: You can appreciate the true elevation of the Peach Orchard and the sharpness of the slope leading up to it.

While the open space of the Wheatfield is some 100 yards behind you, the Stony Hill has become a crucial element of the Wheatfield fighting. This was hard fighting because if a Confederate force were to break through the Wheatfield and the John Weikert Farm, it would be a bad day for the Union. First, the Confederates could reach Cemetery Ridge from the south and start to roll up the "real" flank Lee had intended to hit when he made his plans that morning. Second, they could cut off and leave stranded all of the Union forces at the Peach Orchard and along the

Emmitsburg Road, forcing them to pull out. Such a withdrawal would open the ground and enable McLaws' Division to march up and initiate the "Plan A" attack up the Emmitsburg Road that had been intended all along.

Your hike to this spot brought you past the most well-known of the Irish Brigade monuments with its Celtic Cross formed of green marble. However, the monuments for the 116th Pennsylvania next to you and for the 28th Massachusetts across the street behind you also commemorate units of the Irish Brigade. The Celtic-Cross monument is a memorial to the brigade's three New York regiments. The 28th Massachusetts from Boston had as many men in its ranks as these three New York regiments combined.

(It is said President John F. Kennedy during a battlefield tour in 1963 stopped by the monument with delight because he was able to read the Gaelic phrase: Faugh A Ballaugh. It is pronounced fog-aw-ballick and it means: "Clear the way!")

In the initial positions taken by the Union Third Corps, the 3rd Michigan's long skirmish line proceeded from the point of woods to your left all the way up to the nearest corner of the Peach Orchard. It is evident from this perspective that no single regiment could hold such a line for long.

As the Confederate attack developed, Kershaw came into this position in order to support Brig. Gen. George Anderson's Brigade off some distance to your left and rear. After driving back troops of the Fifth Corps, Kershaw here would confront the counter-attacks of two Union brigades from the Second Corps: Brig Gen. Samuel Zook's brigade coming from your right and Col. Patrick Kelly's storied Irish Brigade coming up behind you. Your walk from Weikert Lane across the Wheatfield and up Sickles Avenue followed the direction of Kelly's advance.

President Kennedy with his entourage on a battlefield tour in March 1963

Hiking Tour Stop #15 (Final): Peach Orchard, 141st Pennsylvania Monument

- **GO:** Follow Sickles Avenue north toward the Wheatfield Road. Turn left on Wheatfield Road and walk 0.25 miles up the slope, using the grass shoulder on the right side of the road. Return to the northeast corner of the Peach Orchard marked off by the asphalt stairs with handrails.
- **ON THIS LEG OF THE HIKE:** You are walking uphill along the Wheatfield Road past the monuments to McGilvery's batteries: 9th Massachusetts, 5th Massachusetts, 10th New York and 15th New York. (The 10th New York placed a separate monument but during the battle was folded into the 5th Massachusetts. The 5th Massachusetts' monument bears the Maltese cross of Fifth Corps but in this battle was part of the army's Artillery Reserve.) You are walking into the charge of the 21st Mississippi headed down this road directly at you.
- **STOP:** at the top of the asphalt steps.
- **LOOK:** toward the Trostle Farm behind you and across the ground you have crossed.

Quick Summary: You can now survey the landscape of your hike from the Emmitsburg Road to your left down to the Trostle Farm, Cemetery Ridge behind it, the Trostle and Weikert Woods to the right of the farm, to Little Round Top and the Wheatfield. You occupy the center point of a wide tract of ground that on July 2nd, 1863, was the amphitheater of the bloodiest four hours fighting of the Civil War.

As the Mississippi troops came plowing through the peach trees behind you, the 141st Pennsylvania in line behind you and to your right opened fire and stopped their advance. These men, commanded by Col. Henry Madill, won precious minutes for the artillery crews in the orchard and on the Wheatfield Road to get their guns out.

Madill

The minutes were paid in blood as the 141st Pennsylvania was almost wiped out. No Union Third Corps regiment suffered a heavier percentage of losses at Gettysburg. The 141st Pennsylvania monument is at your left facing north.

More Detail: The artillery batteries you passed on the last leg of your hike were part of the force that the commander of the 21st Mississippi, Col. Benjamin Humphreys, felt it was necessary to deal with.

If Col. Humphreys had turned his men north with the rest of Barksdale's Mississippi brigade to pursue the second leg of their attack up the Emmitsburg Road toward the Klingel Farm, he would have left these Union batteries plus any number of scattered Union troops in his flank and rear. No competent officer would do such a thing. As was true of so many Confederate commanders in the fighting on July 2nd, the real-time circumstances on the ground forced Col. Humphreys to adapt his tactics. You may conclude your hiking tour by consulting the narrative under Walking Tour Stop #5 on page 11.

Humphreys

MONUMENT PHOTOS without captions

PAGE 21: Louisiana Memorial
PAGE 23: 1st New York Battery G
PAGE 31: 3rd Massachusetts Battery
PAGE 32: 28TH Massachusetts
PAGE 33: 141st Pa.

Tour C: Driving Tour

The Driving Tour features 19 tour stops over 20 miles. It should take about three hours but if you are lingering, reading, taking photos, discussing everything and having a good time, it could take all day.

The Driving Tour begins at the place where the dramatic decision is made by Gen. Robert E. Lee to mass his assault force at the Peach Orchard in order to strike the Union left. The tour then directs you to follow the route of the Confederate divisions on their march to the battlefield, including the controversial "countermarch" that occurs after it is discovered the original route was not safe from observation. Many of the historic roads for this marching are still in place.

The Driving Tour will then bring you onto Gettysburg National Military Park to important vantage points to survey the landscape of the fighting on July 2nd in such a way as to build a deeper appreciation for the central importance of the Peach Orchard.

The Driving Tour is designed exclusively for passenger vehicles. It is no good for tour buses. (Blackhorse Tavern Road, Old Mill Road and Willoughby Run Road are not rated for vehicles over eight tons.) The tour is also not intended for mini-buses or RVs as such vehicles would create a road hazard at many stops.

It would be ideal for the driver to have a companion who can read the **GO** directions while the driver concentrates on the road and traffic. If you must take the Driving Tour alone, read the **GO** directions thoroughly at each Driving Tour Stop before proceeding to the next Driving Tour Stop.

Extra caution is required for Driving Tour Stops #3 through #7 because they stop alongside public roads that are not part of Gettysburg National Military Park and may have residential and commercial traffic. Therefore:

1. You must engage your hazard flashers at these stops.
2. You must be sure that you have left ample room for traffic to pass.
3. You must show respect for private property.

Once you enter the national park at Driving Tour Stop #8, parking on the right side of the one-way tour roads is permitted without the need for hazard flashers. However, it is recommended you do use hazard flashers at Driving Tour Stop #11 on South Confederate Avenue and Driving Tour Stop #12 on Brook Avenue in front of the Brook Avenue sign.

Each Driving Tour Stop has a header followed by four finger-point bullets with specific instructions.

- **ON THIS LEG OF THE DRIVE:** This describes the significance of the road or ground you will cover.
- **GO:** These detailed instructions explain the route you should drive. It includes mile points measuring the distance to landmarks en route. The mile-point measurements start at ZERO after each Driving Tour Stop. The mile-point distances described in each leg are measured from the previous Driving Tour Stop. Mileages are approximate by the odometer. While it is not necessary for you to reset your trip odometer to zero at each stop, it might be helpful to do so.
- **STOP:** This provides specific instructions on where to halt your vehicle safely for the Driving Tour Stop.
- **LOOK:** This tells you where to direct your attention to understand the Driving Tour Stop narrative.

Quick Summary: This will provide a few brief paragraphs to introduce the significance of the place where you have stopped.

More Detail: At most stops this will provide a fuller narrative of events and in a few cases a larger discussion of the Battle of Gettysburg.

QUICKER TOUR: The route of the Driving Tour without any stops at all takes 45 minutes at normal speed limits. Stops where you remain in the vehicle will each require up to five minutes if you read the full narrative. Tour stops where you exit the vehicle may each require from five to 15 minutes if you read the full narrative. As a result, the full Driving Tour may consume at least three hours. To spend less time, the Driving Tour can be abridged by skipping Longstreet's countermarch and a few other Tour Stops. Instructions for the QUICKER TOUR are inserted at the end of the narratives for Driving Tour Stops #2, #8, #11 and #16.

Driving Tour Stop #1:
The Lutheran Seminary

- **ON THIS LEG OF THE DRIVE:** Your trip through town brings you through an important segment of the battlefield often overlooked because it is not a part of the national park nor is it noted on most auto tours. Baltimore Pike and Washington Street were major routes of Union advance and retreat. Chambersburg Street west from the town circle is the important artery in the retreat of the Union First Corps and in the advance of Confederate Col. Abner Perrin's Brigade of South Carolinians to the center of town. At sunrise on July 2nd the townscape you will pass through was firmly in Confederate hands.
- **GO:** From the National Park Service Visitor Center, leave by the Baltimore Pike exit (east) and turn left at the traffic light onto the Baltimore Pike. Proceed through Gettysburg to the traffic circle at the center of town. Entering the circle from the south, go three-quarters (270 degrees) of the way around and leave the circle onto Chambersburg Street heading west.

  Or you can leave National Park Service Visitor Center by the Taneytown Road exit (west) and turn right onto Taneytown Road. Remain straight at the traffic signal onto Washington Street and continue to the intersection with Chambersburg Street. Turn left onto Chambersburg Street heading west.

  Either way, you will now follow Chambersburg Street west past Franklin Street. Note Chambersburg Street (Route 30) bears to the right at the three-way intersection with Springs Avenue and West Street. It also changes names to Buford Avenue.

  As you head out of town on Buford Avenue, the road will rise up a gentle slope. The campus of the Gettysburg Evangelical on top, is the Civil War-era seminary building, presently the home of the Seminary Ridge Museum.

  Turn right at the Stop sign and park in museum parking. Do not park on Seminary Ridge Avenue itself. Cross the street to the Seminary building and walk around the left (north) side to the "back" of the building. (This leads to the ground-floor entrance to the museum.)
- **STOP:** at the left rear of the Seminary Ridge Museum building with a view over the town of Gettysburg.
- **LOOK:** toward the town southeast toward Cemetery Hill.

Quick Summary: At the time of the battle, this spot offered open views of Cemetery Hill and Cemetery Ridge. Somewhere in this vicinity a table would be set up at which Gen. Robert E. Lee explained the attack plan for the Confederate Army of Northern Virginia on July 2nd. Lee issued verbal orders to launch the main attack of the day against the Union left with the assault force taking position in the Peach Orchard.

Lee gave instructions directly to Major Gen. Lafayette McLaws, a division commander, which was an unusual step. Lee told McLaws to place his division in a line perpendicular to the Emmitsburg Road. Ordinarily it would be up to the corps commander, Lt. Gen. James Longstreet, to give such instructions to a division commander like McLaws. Lee decided to circumvent this chain of command. Longstreet responded by giving his own orders to McLaws that countermanded Lee's instructions.

Lee stepped forward again and reiterated his original wishes, overruling Longstreet. The army commander by superseding the corps commander's judgment had relegated Longstreet to a secondary role concerning the initial maneuvers of the day.

More Detail: There are some evergreen trees here, which are not native to this region. Photos from 1863 show hardwood trees, which are much easier for climbing. In several accounts, observers found a good perch from which to get a better view over the field – something you could not do in these evergreens. In one account, a tree had fallen, its trunk providing a handy bench.

The area around the 1832 Seminary building provided a more open view over the town and the battlefield than exists today, even in winter. The photo on the previous page may help you discern Culp's Hill and Wolfs Hill. A sharp eye will spot the statue perched atop the New York State Memorial in the national cemetery, which then orients you to the location of Cemetery Hill. It may also be possible to spot the green Gettysburg water tower that marks the eastern edge of Cemetery Hill, clear to see in winter but more difficult when the trees have filled in.

At the time of the battle, this spot offered open views of all of those landmark features, including Cemetery Ridge and to a limited extent the Emmitsburg Road. If you visit the museum and pay extra for the twice-daily cupola tour ($29 at the time of this writing), the vantage point from five stories above will enable you to see many of the landscape features described. (You will also be supporting a most worthy institution.) Much of the landscape visible from the cupola would have been visible from ground level in 1863. However, it is not essential for the purposes of this tour to go into the cupola or the museum. The point is that as long as you are making the effort to seek out landmarks, you are looking in the correct general direction.

Lee

Somewhere on this slope in front of the Seminary, a rider came up alone in search of the army commander. He dismounted and the two men, good friends and stalwart comrades in arms, stood together and contemplated Cemetery Hill, the Union position and the events of July 1st. Longstreet's men were still many hours behind him, stuck in traffic in the Cashtown Pass or lumbering up the dusty Chambersburg Pike, but here on this spot Longstreet offered his opinion to Lee that fighting in Gettysburg had been a mistake. He then opined further that Lee should move the army south to a better defensive position in Maryland and closer to Washington. It was an opinion that was offered in error because Longstreet was ignorant of the country and of the road systems over which he was recommending such as difficult maneuver. Lee knew better than Longstreet that such a movement in these conditions was implausible.

The next morning, July 2nd, it is around this building somewhere, likely on the left (north) side or in front by the road, that the headquarters staff of the Army of Northern Virginia set up a table. This simple table was the setting for a drama that would set the stage for the events of July 2nd and determine the outcome of the battle – perhaps of the war and still further the future of the nation.

It was at this table, within a few dozen yards of where you are standing, that Gen. Lee issued verbal orders to launch the main attack of the day from the Peach Orchard against the Union left on the Emmitsburg Road and on Cemetery Ridge. The objective was to drive in the Union flank and to strike the fatal blow against the "decisive point" of the Union defense: Cemetery Hill. But this discussion would also engender a conflict in authority that would cause Longstreet to step back from direct command during the crucial first hours of maneuver.

In the infantry, then and now, the army commander gives orders to his corps commanders. The corps commanders exercise control of their various divisions.

Having observed the landscape here and on Culp's Hill, and having received his scout's report that the ground around Little Round Top was open, Lee announced his intentions. He did so while standing at this table. It produced an odd and awkward moment. Lee decided to bypass Longstreet as First Corps commander and to give orders directly to one of his division commanders, McLaws. Lee instructed McLaws to move his division to the Peach Orchard and to form it in a line perpendicular (at right angles) to the Emmitsburg Road. He used his finger to indicate the

> **What does "decisive point" mean?**
>
> It is the notion, evolved from Napoleonic tactics, that every defensive position has one critical point that – if detected and struck with sufficient force – will break the defending army and bring victory to the attacking army.
>
> Since it can decide the battle, it comes to be known as the **"decisive"** point. This idea was articulated in the writings of Antoine Henri Jomini, a rigorous part of the West Point curriculum.

desired position on a map. Lee wanted the men to deploy facing northeast toward Cemetery Hill.

Standing nearby, Longstreet judged it was his prerogative as corps commander to be giving such specific instructions concerning the deployment of his divisions. He stepped up to the table and gave his own orders to McLaws that contradicted those of the army commander. Longstreet used his finger to indicate the position on that same map. He instructed McLaws to move his division to the Warfield Ridge and to form it in a line facing the Peach Orchard and parallel to the Emmitsburg Road. Longstreet wanted the men to deploy facing east toward the Peach Orchard itself.

This was at odds with what Lee had instructed.

With this, Lee pointed to the map again and reiterated his initial instructions, overruling Longstreet. McLaws would say in post-war accounts that the moment was awkward in the extreme. He claimed this caused Longstreet to withdraw and to spend much of the day sulking. McLaws would ultimately denounce Longstreet, his fellow Georgian and one-time benefactor, as "a humbug, a man of small capacity, very obstinate, not at all chivalrous, exceedingly conceited and totally selfish."

Those less inclined to denounce Longstreet nevertheless do see this as a moment in which Lee, by superseding Longstreet's judgment, relegated him to a secondary role concerning the initial maneuvers of the day. Longstreet would write in his 1896 memoir, "I was relieved for the time from the march." Thus it was to be up to Lee's appointed scout, Capt. Samuel Johnston, the engineer who had scouted out Little Round Top, to guide the First Corps divisions to the battlefield while Longstreet would ride a mile or more back on the column of march.

That is until everything went wrong.

Longstreet's First Corps had three divisions: McLaws', Major Gen. John Bell Hood's Division and Major Gen. George E. Pickett's Division. Hood's Division was to follow McLaws to the Peach Orchard and support him in his attack. Pickett's Division, the smallest of the three, was still back in Chambersburg, assigned to guard the army's supply wagon train and the rear of the army.

Longstreet pointed out to Lee that only three of Hood's four brigades had arrived. Still many miles away on the march was the Alabama Brigade, the largest in Hood's division. It's commander, Evander Law, also happened to be Hood's second-in-command as his senior brigadier. Lee assented to Longstreet's request to postpone the movement until Law's men could arrive and join the others.

When the march to battle began at around noon, Longstreet rode with Hood, who was leading the second division in the column. McLaws rode far ahead of them at the front of the column with the guide, Capt. Johnston. These "small" details would prove to be massive in giving shape to the coming battle on July 2nd, all of it centered initially around that single, critical point on the map: The Peach Orchard. It was at the table near this building where the fateful die was cast.

Driving Tour Stop #2:
Thompson House/"Lee's Headquarters"

- **ON THIS LEG OF THE DRIVE:** You will re-cross the Seminary Ridge line that formed the crux of the Union's final defense on July 1st prior to their withdrawal through town to Cemetery Hill. Seminary Ridge was Lee's base on July 2nd. The War Department when it placed headquarters markers had evidence to show Lee made his headquarters in the orchard seen to your left in rear of the white house at the corner, which seminary students know as North Hall.
- **GO:** Leaving the Seminary Ridge Museum parking lot, make a left back onto Seminary Ridge Avenue and go out the way you came in. Pass the Seminary Ridge Museum building and continue to the Stop sign on Buford Avenue/Chambersburg Pike at mile point 0.2. The Thompson house is directly in front of you across the street. Carefully cross the street and enter the parking lot for "Lee's Headquarters," the Mary Thompson House.
- **STOP:** park and walk to the set of wayside markers to the right (north) of the Thompson House.
- **LOOK:** east toward the center of town.

Quick Summary: The Mary Thompson House at the time of this writing had been recently preserved and restored by the American Battlefield Trust. It is known to generations of visitors as "Lee's Headquarters," a privately owned tourist attraction.

Although the War Department installed a vertical-cannon headquarters marker for Lee in the orchard across the street, the Thompson house site you are visiting was likely used by Confederate army staff and perhaps by Lee himself to rest or take a meal. The house was occupied in 1863 by the tenant Mary Thompson, an aging widow. Her landlord was Thaddeus Stevens, a prominent leader of the abolitionist "radical Republicans" in Congress.

The accompanying photograph, which shows the landscape features of the town in 1863, was taken from here. It illustrates how the buildings and streets of the town represented an obstacle to an infantry force.

More Detail: While the evidence shows the issuance of orders occurred around a meeting table by Driving Tour Stop #1, it is appropriate to spend a moment in these surroundings as well to further contemplate the command decisions made on the morning of July 2nd.

It was possible on July 1st for squads and companies to pursue retreating Union soldiers through streets and alleys but it was impossible to form brigades or divisions within the town for a major frontal assault on July 2nd. The town is an obstacle for any attack against Cemetery Hill. It was evident, perhaps inevitable, that a continued offensive of any size would have to go around to the right.

Consider for a moment the gravity of the decisions Lee made on Seminary Ridge, committing his "invincible" army to a risky battle.

Turn around and look toward the west.

You are standing on the ground of heavy fighting on the afternoon of July 1. You are looking over a landscape that was littered with destroyed equipment and with appalling numbers of Confederate dead. The frontal assault against the line of Union guns on Seminary Ridge had cost many lives. The monument for 4th U.S. Battery B can be seen in front of you on the western edge of the Thompson property. This was one of the Napoleon batteries blasting canister point-blank into the ranks of Brig. Gen. Alfred Scales' North Carolina Brigade. Another battery commander, Capt. Greenlief Stevens, whose Napoleons were across the street in the seminary, would report, "The whole line of battle from right to left was one continuous blaze of fire."

Consider Lee's frame of mind on July 1st. He could ride forward from Herr Ridge with pride in the victory of his army on the first day of battle in Pennsylvania. They had brought the fight not to slave-holding borderlands like Maryland or Kentucky but to a truly northern state, into the beating heart of the Union. Yet the scenes of death and suffering Lee would have witnessed along his ride would likely have tempered any high spirits.

In this, the third summer of the war, the foe of course remained Lincoln the Oppressor and his invading Union armies. But the real enemy had become war itself. The novelist Michael Shaara a century later would imagine these thoughts in the mind of the great general: "We are adrift in a sea of blood and I want it to end. I want this to be the final battle."

Now look farther west beyond Herr Ridge to the horizon. In most weather conditions you should be able to see the blue wall of South Mountain 12 miles to the west. It is in fact not a mountain but a ridge that runs north-south all the way to Virginia, where it continues south to Georgia in the form of the famous Blue Ridge. South Mountain was the major geographic feature – and impediment – that Lee would have to deal with as he moved his troops toward Gettysburg. Remember this view as we discuss South Mountain in more detail at Driving Tour Stop #5 – the Black Horse Tavern/Fairfield Road.

For the time being, consider the job of the army commander as you stand here at the site labeled "Lee's Headquarters."

Lee had already chosen to disregard the advice of Longstreet, his senior lieutenant and confidant, who had urged him to quit Gettysburg and to redeploy to a new defensive line somewhere south. (It is difficult to calculate or imagine what network of roads Longstreet might have thought existed to make such a risky move possible.)

Lee's job as army commander was to consider all factors, including the views of his subordinates, and then to be decisive.

For the soldier on the line, the fatal defect is cowardice. For field officers, the fatal defect is hesitation. For the commander of the army, the fatal defect is indecision. The only consensus Lee was required to achieve was within his own mind. Lee was not obliged to explain to Longstreet or anyone else why he would dismiss his suggestions. Now on the day of battle, Lee had chosen to intrude on Longstreet's prerogative by issuing orders directly to one of his subordinate commanders.

Much is made in postwar writings and statements that Longstreet was at odds with Lee during the Pennsylvania Campaign and so may have subverted his plans. Most of this verbiage concerning a rift between these two men is imaginary.

Lee had sound reasons for every judgment he made on most days of the war, including on this day. The fault lies with scholars and critics who expect perfection. It is unattainable. Lee had not made his decision in a fit of rashness. He had made his decision with care. His conclusions were supportable. It just so happens his decision was wrong.

QUICKER TOUR: Skip Driving Tour Stops #3 through #7 (Longstreet's countermarch). Proceed directly to Driving Tour Stop #8 Warfield Ridge War Dept. Observation Tower.

Leaving the Lee's Headquarters parking lot, carefully cross Chambersburg Pike back onto Seminary Ridge Avenue. Continue straight past Driving Tour Stop #1 "The Lutheran Seminary" and go straight to the traffic light. Cross Fairfield Road onto West Confederate Avenue. Go straight for two miles to the Stop sign at Millerstown Road. Continue straight across Millerstown Road and proceed to the parking lot on the left for the Warfield Ridge War Dept. observation tower. Pick up the tour instructions for Driving Tour Stop #8.

Driving Tour Stop #3: Herr Ridge

☛ **ON THIS LEG OF THE DRIVE:** You will cross the main Union battle line on July 1st along McPherson Ridge, marked by Union regimental monuments as far as the eye can see in both directions. You will end up on Herr Ridge, the main Confederate battle line on the morning of July 1st. Getting there you will be traveling on the Chambersburg Pike, the principal road used by Confederate forces on their approach to the battlefield. The town of Gettysburg is famous as the intersection of 10 major roads, but seven of Robert E. Lee's nine infantry divisions would use this one road to get to the battlefield.

☛ **GO:** Leaving the parking lot for Lee's Headquarters (Thompson House), make a right onto Chambersburg Pike and head west. Pass the West End Guide Station on McPherson's Ridge at mile point 0.4 and cross Willoughby Run, the national park boundary, at mile point 0.7. You will reach historic Herr Tavern and the Herr Ridge Road at mile point 1.1. The intersection has a pair of traffic lights. Turn left at the second light onto historic Herr Ridge Road.

☛ **STOP:** park on the right side of Herr Ridge Road along the gravel shoulder opposite the tavern. Engage your hazard flashers. Remain inside your vehicle.

☛ **LOOK:** Straight ahead on the road.

Quick Summary: You are on the ridge on which Confederate forces of the Third Corps formed their main battle lines on the morning of July 1st before advancing against the Union positions on McPherson's Ridge, which you crossed on your drive.

You are parked on the historic road that all or part of McLaws' First Corps division used as the men arrived in the dark to move to the army's southern flank. After McLaws moved through, Hood would bring his division onto this part of Herr Ridge Road and go into camp in the fields and woods around it.

This is the first of several historic roads on the driving tour. The roads are historic in that they generally follow the 1863 farm lanes that were here at the time of the battle (allowing for engineering modifications in grade and angle that would have been made during modern paving.)

More Detail: The area you are about to explore is a quiet and unspoiled region of the battlefield on roads that seem remote from the national park. You are parked about half a mile from the park's official western boundary. Most visitors to Gettysburg will see the Chambersburg Pike at some point but few of them experience these rural pathways. Yet these were the critical routes of Longstreet's march and countermarch. Understanding those events is essential in setting up the action on July 2nd – and understanding the action on July 2nd is essential to any real grasp of the Battle of Gettysburg.

The roads on the next few legs of the tour pass through open country with beautiful pastures and farms. However, there is one important road, the Forney Farm lane, which was used on the march but which no longer exists. We will see the points where this long-lost dirt lane started and ended.

McLaws made use of Herr Ridge Road and the Forney Farm lane for at least a part of his force. The movement of Hood's Division, the next one up in Longstreet's Corps, is less clear. All of the officers from brigade level on up began their Gettysburg reports with the start of the fighting without any discussion of their march into position.

Hood would write that when he reported to Lee on Seminary Ridge (Driving Tour Stop #1) on the morning of July 2nd, he came up with his division, which suggests they had broken camp and marched up the Chambersburg Pike with him. It is likely Hood may not have meant it quite that literally and that his men remained on Herr Ridge. Longstreet in his report and memoir would say merely that Hood's troops were "close at hand."

Hood

Driving Tour Stop #4:
Forney Farm Lane Location

☛ **ON THIS LEG OF THE DRIVE:** Your route will follow the line of march of Longstreet's two divisions, McLaws followed by Hood. After crossing historic Old Mill Road, go another third of a mile. You will pass farm fields to your right. Somewhere across these fields was a farm lane. McLaws used this 1863 lane to get to the Black Horse Tavern.

☛ **GO:** Drive straight on Herr Ridge Road away from the Chambersburg Pike behind you. Cross the intersection with Old Mill Road at mile point 1.0 and Kestrel Drive at mile point 1.1 You will come to a private black mailbox on the right at mile point 1.3. Pull off onto the grass and gravel shoulder just beyond the mailbox.

☛ **STOP:** park on the right side of Herr Ridge Road along the shoulder where you can do so safely. Engage your hazard flashers. Remain inside your vehicle. (This is private property but the shoulder is a public right-of-way.)

☛ **LOOK:** to your right across the open farm fields.

Quick Summary: We directed you here so you can imagine where a primitive farm lane may have run in 1863. It intersected Herr Ridge Road somewhere around here, most likely somewhere behind you, and continued across the fields into the wood line you can now see to your right (west). Beyond the woods, the farm lane would have turned sharply left across the Forney property to reach the Blackhorse Tavern Road.

This long-lost farm lane is the route McLaws followed. His men went into camp all around here at about midnight on July 1st, possibly extending a half mile west of here to Marsh Creek in order to water their animals. On July 2nd, Hood's Division would follow McLaws on this lane. Many of Hood's men were still on this lane and on the open ground to your right when the whole column came to a halt and the countermarch backward would soon commence.

We will now take a more circuitous route on paved roads to get to the other end of the Forney Farm lane shortcut to see the intersection with Blackhorse Tavern Road where McLaws was headed.

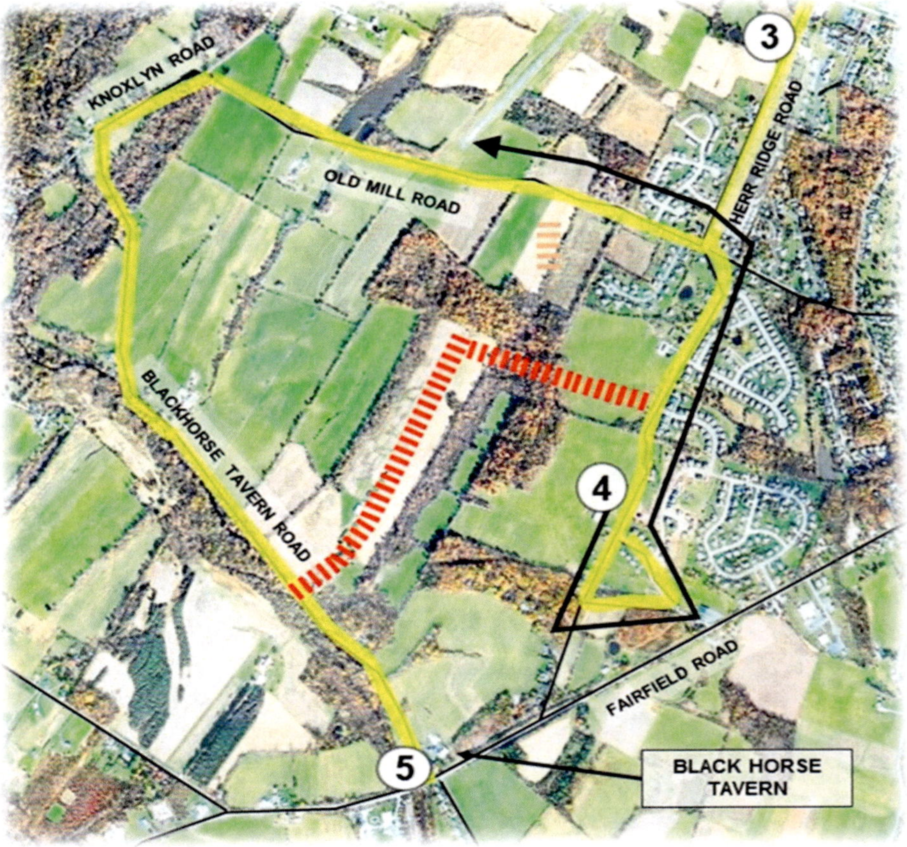

The historic 1863 Forney farm lane no longer exists but it is thought to have followed the red track. After leaving Driving Stop #4, you will turn left twice on Fairplay Road to return to Herr Ridge Road, then drive back to the intersection for a left turn onto Old Mill Road

Driving Tour Stop #5: Black Horse Tavern/Fairfield Road

☛**ON THIS LEG OF THE DRIVE:** Your drive will take you past the historic Adam Butt Farmhouse, where there may have been a second farm lane in 1863 also used by the Confederates. The drive will then bring you back to Old Mill Road and down the line of Longstreet's artillery battalions along Old Mill Road. You will follow historic Knoxlyn Road around the back of his marching column. The point where Knoxlyn Road intersects Blackhorse Tavern Road is the site of an old mill on the creek. Some of McLaws' troops may have camped down here and approached Black Horse Tavern from the mill site.

☛**GO:** Continue straight on Herr Ridge Road. Pass the first intersection with modern Fairplay Road. (You will pass the historic Adam Butt Farmhouse on your right.) Next you will come to a Stop Sign at the second intersection with Fairplay Road at mile point 0.3. Turn left onto Fairplay Road. You will come to another Stop Sign at mile point 0.4. Turn left again and return to Herr Ridge Road at mile point 0.6. Turn right onto Herr Ridge Road to pass your previous Driving Tour Stop #4 "Forney Farm Lane Location." Return to historic Old Mill Road at mile point 1.3.

Turn left onto Old Mill Road. When you hit the "T" intersection at mile point 2.6, turn left onto historic Knoxlyn Road. Knoxlyn Road reaches the intersection with historic Blackhorse Tavern Road at mile point 2.9 in front of a modern bridge that crosses historic Marsh Creek. (The old mill was here.) Do not cross the bridge. Turn left onto the Blackhorse Tavern Road. Follow Blackhorse Tavern Road with Marsh Creek on your right. At mile point 4.1 you will pass a modern white farm complex on your left. This approximates the location of the lower segment of the Forney Farm lane in 1863. (This is private property.)

Continue on Blackhorse Tavern Road. (The next white farmhouse on the left is the original Mark Forney Farm.) You will reach the Fairfield Road and the historic Black Horse Tavern building on your left at mile point 4.0. ☛ **STOP:** park on the right side grass shoulder of the Blackhorse Tavern road. Engage your hazard flashers. Exit your vehicle, cross the Blackhorse Tavern Road and walk over to the iron signs in front of the Black Horse Tavern. (The historic building is on private property. Do not trespass. Confine your visit to the signs on the shoulder of the Fairfield Road.)

☛**LOOK:** Turn your back to the iron signs and look left and right up and down the Fairfield Road. To your right is the bridge crossing Marsh Creek. Across the street you can see where the Blackhorse Tavern Road continues south. The ridge of Bream's Hill is sloped sharply above the left side of the road.

Quick Summary: The late hours of July 1st witnessed a high degree of activity and anxiety in this area and at the Black Horse Tavern, an enterprise of Francis Bream in 1863.

First, a brigade of Alabama troops took position at the former mill site at Marsh Creek, which you just passed, and also on Bream's Hill, the rise of ground to the left (east) of your present location on Fairfield Road. They set up guns facing east and south while deploying lines of skirmishers everywhere.

Hours after dark, a second drama unfolded as a dusty cadre of Union officers led by Brig. Gen. Andrew Humphreys stepped out of the darkness into the parlor of the Black Horse Tavern. The proprietor warned these Union officers that they were taking a big risk because Confederate skirmishers were in the woods right behind the house. Humphreys back-pedaled delicately and slipped back up the road from whence he had come.

Black Horse Tavern

Hours later, McLaws' Division came in from the Chambersburg Pike. They backfilled the perimeter of the army's southern flank as McLaws made his headquarters somewhere up on Herr Ridge. (One has to wonder if Mr. Bream informed any of the Confederates that a Union general had stood in his parlor just a few hours earlier.)

More Detail: Heavy fighting near Gettysburg had been evident throughout the afternoon of July 1st, spilling over onto Fairfield Road 1.5 miles east of this spot. The noise of battle would have added drama to the waves of smoke rising on the eastern horizon. The evening hours brought the arrival of the Alabama Brigade under Brig. Gen. Cadmus Wilcox. Wilcox's Brigade was part of Major Gen. Richard Anderson's Division.

A. Humphreys

Each of the three Confederate corps had three divisions. Anderson's Division was in Third Corps. Two divisions of the Third Corps had arrived first on July 1st and had carried the bloody weight of the fighting. Now the last of the three divisions, Anderson's, came up behind them from the west to secure the line facing Gettysburg.

Anderson had to pay special attention to his far right, which was the southern flank of the whole army. He assigned it to Wilcox. The Alabama general used either the Knoxlyn Road or the Old Mill Road to get to this right-flank position.

Union Gen. Humphreys was marching north from Emmitsburg with two brigades at this time on nearby side roads. He was being guided by a division staff officer who insisted the march to the Black Horse Tavern was safe and conformed with his orders. The 52-year-old soldier's instincts kicked in. Humphreys paused the night march and rode ahead to check out the place for himself, concerned he was leading his men in the wrong direction and was bringing them close to danger. He was correct on both counts.

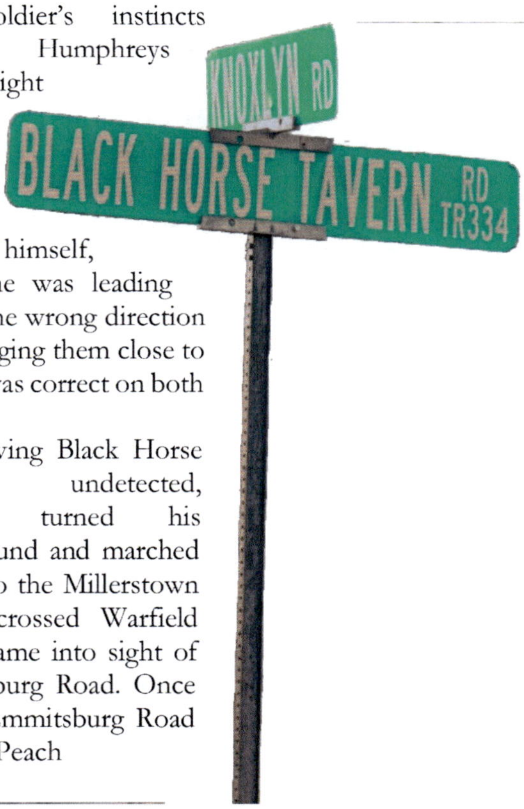

After leaving Black Horse Tavern undetected, Humphreys turned his brigades around and marched them back to the Millerstown Road. He crossed Warfield Ridge and came into sight of the Emmitsburg Road. Once across the Emmitsburg Road at the Peach

Orchard, he found the camps of the Union army.

Let us take some time out from our tour narrative to think about the overall Gettysburg Campaign.

First, consider the road on which you are standing: the Fairfield Road, so important yet so often overlooked. (On many 19th Century maps it was referred to as the Hagerstown Road.)

Consider the geography.

To our west rises the South Mountain ridge. Hopefully (weather permitting) you got a good look at this high ridge when you visited Driving Tour Stop #2 – Thompson House. As was noted at that tour stop, South Mountain is the dominant geographic barrier that affected the movement to battle by the Army of Northern Virginia.

On Thursday, June 25, Lee reached Chambersburg on the western side of South Mountain. Gettysburg is on the eastern side. The distance between is 25 miles.

Every student of the battle knows the opposing armies came to grips at Gettysburg because of the 10 major roads leading into town. The writer Shelby Foote referred to the town as having 10 roads "reaching for trouble in all directions."

Forget this for the moment.

We are now considering only South Mountain. The notion of 10 roads does not apply. Troops could cross South Mountain on only two roads. These went through the two gaps: the Cashtown Pass and the Monterey Pass.

The more famous gap is the Cashtown Pass. It lies between Chambersburg and Gettysburg. It carried through it one road, the Chambersburg Pike. You went up the Chambersburg Pike when you left Driving Tour Stop #2 – Thompson House.

Less famous is the Monterey Pass.

Standing as you are on the Fairfield Road, if you were to drive west from here you would reach the village of Fairfield in five miles. Beyond Fairfield, one of two rural tracks, the Jacks Mountain Road or the Iron Springs Road, would bring you up the slope of South Mountain into the Monterey Pass.

The Monterey Pass in 1863 carried through it one road, the Old Waynesboro Pike (present-day Charmian Road). Once you got through the pass and down the other side of the mountain you would come to the town of Waynesboro.

Consider the decisions Lee made as he brought his army north into Pennsylvania. As was his custom, Lee led his army with Second Corps, the legendary quick-marchers who had so recently been under the command of the late Stonewall Jackson. Now they were under the command of Lt. Gen. Richard Ewell, the man Jackson had recommended as his replacement.

Ewell had moved to Carlisle with two divisions. His eye was fixed on Harrisburg, the state capitol. A third division under Major Gen. Jubal Early had gone through Gettysburg and York to Wrightsville on the Susquehanna River. Farther south, cavalry brigades had skirted Washington. Confederate troops were showing up everywhere in force.

This aggressive dispersal presented quite a challenge to any Union army commander tasked with confronting these invaders while also positioning his army to protect Baltimore and Washington.

The stress proved too much for Major Gen. Joe Hooker, who was sending messages to the War Department that alarmed the Lincoln Administration. They took action – but not the action Hooker had expected. Shortly before dawn on Sunday, June 28, orders arrived to relieve Hooker of command. Major Gen. George Gordon Meade, commander of Union Fifth Corps, was elevated to command of the whole army.

Wilcox

Lee in his deployment of Second Corps under Ewell had set the stage for the opposing Union commanders to make errors. Lee had done this so often with foxlike ingenuity. But had Lee avoided making mistakes of his own? Had he deployed his own forces correctly?

Consider this question: Why did Lee in his advance to Gettysburg not make use of this Fairfield Road alongside which you are now standing?

Many civilian historians dismiss "what-if" speculations as a waste of time (although it is common at military war colleges for professionals to delve into exactly such scenarios). Let us indulge for a moment. Let us contemplate the importance of Fairfield Road as a missed opportunity.

Lee arrived in Chambersburg with the Third Corps, commanded by Lt. Gen. A.P. Hill. Bringing up the rear of the army was Longstreet's First Corps. As Harrisburg emerged as a potential target of their campaign, Lee ordered Longstreet's men up to join him in the vicinity of Chambersburg. As a result, by Saturday, June 27, Lee had two corps with all six of their divisions clustered in and around Chambersburg.

Officers like Robert E. Lee and his contemporaries were educated in the maxims of Napoleon as found in the writings of Antoine Henri Jomini. One famous maxim was to keep your corps within supporting distance of one another.

How about ideas from other continents and cultures? Transcendentalism was quite the fashion in literary circles at the time but had little influence on military thinking. Allow us to transcend these limits of Western thinking and venture back to China of the 5th Century BC. One general of this ancient period, Sun Tzu, is often paraphrased to have said: "Every battle is won or lost before it is ever fought." (Sun Tzu never actually wrote this phrase explicitly but it is implied in a number of his other sayings.)

Jomini

Can this saying be applied to Lee?

With Ewell's Second Corps spread out from Carlisle to Wrightsville, Lee had concentrated his remaining two corps (six divisions) in Chambersburg. In the event of a sudden need to move due east toward Gettysburg, Lee had limited himself to one gap, the Cashtown Pass, and to the one road going through it, the Chambersburg Pike.

Sun Tzu might instruct us that with this single error, Lee had lost the coming battle.

Let us imagine for a moment that Lee had not ordered Longstreet's men up to Chambersburg. Let us imagine for a moment Longstreet's First Corps had remained back near Waynesboro. Once Lee decided he would move all of his forces east toward Gettysburg, Longstreet's men could have used the Monterey Pass and come through Fairfield. His divisions on July 1st could have been marching up the Fairfield Road on which you are now standing.

Let us return to what actually happened.

On June 30th and July 1st, during the day and overnight, Lee had six divisions moving eastward on a single road, the Chambersburg Pike. The first up were two divisions of Third Corps. On July 1st they slammed into the Union defenses on McPherson's Ridge and Seminary Ridge.

Chinese Terracotta Warrior general

Coming from the north, two divisions of Ewell's Second Corps approached Gettysburg on the only two available roads, one from Carlisle and one from Heidlersburg. Ewell had to send his third division under Major Gen. Edward Johnson all the way back to Chambersburg to use the Cashtown Pass and the Chambersburg Pike for his approach to Gettysburg.

This movement by Johnson's Division created a tremendous roadblock made worse by the fact Johnson was bringing along the entire wagon train for the Second Corps. This traffic jam blocked Longstreet's First Corps from crossing through the Cashtown Pass until almost nightfall on July 1st.

Imagine how different the battle on McPherson's Ridge and Seminary Ridge might have been on July 1st had Longstreet's two divisions arrived early in the day coming up the Fairfield Road, approaching the battlefield in tandem with the troops of Hill's Third Corps rather than 12 hours behind them. Imagine at the height of the fighting on Seminary Ridge if McLaws' Division had come pounding into the Union left flank, followed within the hour by Hood's division. Think of the difference this might have made in Confederate hopes to capture Cemetery Hill, depriving Meade of the road junction he needed to concentrate his army.

Once Longstreet's divisions moved north to Chambersburg, the use of the Monterey Pass and Fairfield Road became a lost option. Longstreet would have needed a full day just to move his divisions back south to Waynesboro before even turning toward Gettysburg. It could not be done.

Lee lost the Battle of Gettysburg. That is plain enough. When did he lose it? With what decision, with what mistake, did Lee set up the defeat of his vaunted army?

The Fairfield Road in front of you might suggest this answer: Lee lost the battle when he moved Longstreet north of the Monterey Pass and in doing so gave up the use of the Fairfield Road. Lee made ample use of the Fairfield Road on the way out on July 5th during his retreat, having been cheated of victory, but he failed to make use of it on the way in on July 1st, when victory was still attainable.

Driving Tour Stop #6:
Breams' Hill Countermarch Point

- **ON THIS LEG OF THE DRIVE:** Your drive retraces the final movement of McLaws' Division, led by Capt. Samuel Johnston, along Blackhorse Tavern Road under the protection of the ridge along your left, which is an extension of Bream's Hill. As the road bends left and climbs over the ridge, you share in the experience from 1863 of the wide open landscape suddenly emerging into view for miles ahead of you.
- **GO:** Carefully cross the Fairfield Road and continue to follow the Blackhorse Tavern Road heading south and then southeast. Pass the intersection with Plank Road at mile point 0.2. Follow Blackhorse Tavern Road up the incline east over Breams' Hill. Cross the top of the ridge at mile point 0.3 to come into full view of the open fields and distant tree line denoting Warfield Ridge. You might be able to see the top of the War Dept. observation tower on Warfield Ridge. If so, pull onto the grassy shoulder just beyond the private shed on the right. If trees block your view of the tower, move up 100 yards and turn left into the white gravel driveway.
- **STOP:** as soon as you have pulled off of Blackhorse Tavern Road and into the driveway. This is private property so DO NOT go up the driveway any farther. (It leads to a house a quarter mile ahead to the right.) Engage your hazard flashers. Remain inside your vehicle.
- **LOOK:** to your right and locate the top of the War Dept. Tower on Warfield Ridge.

Quick Summary: As you crossed Fairfield Road and proceeded up Blackhorse Tavern Road for your first 400 yards, the small but steep ridge on your left obstructed your view to the east. In 1863, this ridge obstructed the view of McLaws and his troops coming up this road. It also concealed them.

But now that you have crested the top of the ridge, the view opens up wide to reveal a mile of rolling fields. The woods in the distance rise up to form the horizon in front of you featuring the peaked roof and observation platform of the War Dept. Tower. That is Warfield Ridge.

The line of woods on Warfield Ridge was much lower in 1863 and may not have been there at all. You would have had a clear view of Little Round Top from here. (When the tower was erected 33 years later there were no trees at all.) From your present-day vantage, the War Dept. Tower is directly in line with Little Round Top one mile behind it.

Men of the Union army signal corps occupied Little Round Top at that moment. So the crucial point here is this: If you can see the Union observers on Little Round Top, it means they can see you.

McLaws knew immediately the march of his division to the battlefield was no longer safe from observation. McLaws brought his entire column to a shuddering halt and summoned Longstreet to the front. Longstreet would order the column to reverse its march back to Herr Ridge – a "countermarch" – to find a different route to the Peach Orchard.

More Detail: The line of woods on Warfield Ridge is made up of Pitzer Woods to the left of the War Dept. Tower and the Biesecker Woods to the right.

Men of the Union army signal corps on Little Round Top were using telescopes and binoculars to scan the countryside and using their red-and-white signal flags to wigwag messages to Meade's headquarters a mile north of them.

McLaws knew the over-arching mandate was to make the approach to the Peach Orchard without being observed. This was no longer possible so he stopped and sought the judgment of higher command.

Before McLaws' hit the crest of Bream's Hill, artillery commander Col. Edward Porter Alexander had led his battalion of six artillery batteries up this same route and had faced this exact same predicament at this exact same spot. Alexander, a gifted and experienced engineer with a keen eye for terrain, made the decision to move up a farm lane marked by the white-gravel driveway on which you may be parked. The

Breams Hill Ridge at Blackhorse Tavern Road

driveway today leads to low ground and a farm pond that was not there in 1863, suggesting the terrain there was low. This route may have concealed Alexander's movement off of the Blackhorse Tavern Road and to the left. Perhaps part of Pitzer Woods, then as now, was tall enough to conceal Alexander's movements.

(Much is made in some circles of the idea that Alexander zigged right along Plank Road before zagging back to the left, or that he found some other way along Marsh Creek as far south as Sachs Covered Bridge. These are engaging topics of discussion. The problem is that Alexander, who was a prolific and accomplished writer, explained his actions in few words in his own Official Report. Faced with the Bream's Hill predicament, Alexander said simply that he went left.)

Consider an interesting point of conjecture. There was one man in Alexander's battalion whose judgment could have made a difference. One of Alexander's batteries included Ordnance Sgt. Henry Wentz, who might have been able to offer timely advice. As it happens, Henry Wentz grew up in Gettysburg. The Wentz family were tenants who occupied the farm right across the street from the Peach Orchard. Wentz would have been familiar with this countryside. Consider also that Alexander in his extensive writing – reports, articles and a full-length history of the war – never mentions Wentz. Nor does Wentz ever write about it in any surviving document. But still.

Whatever it was that Alexander did, McLaws and his guide Johnston did not think to do. Instead, McLaws sent a rider back and waited in place for Longstreet to make his way up to the front of the column. While no report or memoir is known which might verify the specifics of the conversation, we expect that an irritated Longstreet proceeded to question Johnston about this choice of routes. Johnston would have replied to Longstreet that this was not the road he had scouted in his reconnaissance that morning.

At this moment, Longstreet took back command of his troops. Lee had overruled Longstreet's prerogatives, standing over that table on Seminary Ridge at Driving Tour Stop #1, so Longstreet had left it up to Lee's personally appointed guide to lead his troops to battle. Now that the selected route was deemed unsatisfactory, Longstreet took over and immediately ordered a countermarch back to their starting point on Herr Ridge Road in search of a better route.

Further, Longstreet was determined to maintain the order of march that Lee had envisioned. It might have been easier to simply have Hood turn his division around and lead the countermarch. But Longstreet, possibly at McLaws' insistence, sent riders back informing Hood to stand off his division and to allow McLaws' Division to pass him on the countermarch and to resume the lead position. This not only made for some very crowded country lanes, it would have added to the delay already caused by this mistake in the route of march.

Driving Tour Stop #7:
Willoughby Run Road/Blackhorse Tavern Road

☛ **ON THIS LEG OF THE DRIVE:** Your route over the next 4 miles retraces the countermarch of McLaws' Division as it turns around and moves past Hood's Division to resume the lead back to Herr Ridge Road. The marching route includes the same Forney Farm lane they used to get here but which no longer exists.

Your route will bring you from Herr Ridge Road and Old Mill Road back to the Fairfield Road using modern Park Avenue, which runs parallel with the actual route used in 1863. The route then crosses onto historic Willoughby Run Road.

☛ **GO:** Carefully back out of the white gravel driveway onto Blackhorse Tavern Road to make a U-turn. Drive back the way you came down over the ridge and return to the Fairfield Road at mile point 0.3. At the stop sign on Fairfield Road, you will see Blackhorse Tavern Road across the street to your left, which is the road from which you came. You will also see the sharply angled Breams Hill Road to your right. That's where you are going.

When traffic is clear in BOTH directions, turn right onto Fairfield Road and immediately turn left to cross Fairfield Road onto Bream's Hill Road. Go up the hill to mile point 0.6. Turn left onto Herr Ridge Road. Pass the Adam Butt Farm on your left at mile point 0.9 and pass Driving Tour Stop #4 "Forney Farm Lane Location" at mile point 1.1. Turn right onto Old Mill Road at mile point 1.75.

Turn right onto Park Avenue at mile point 2.1. (Park Avenue is a suburban area so be mindful of cars pulling out of driveways, people walking their dogs and children riding their bikes.) Continue to the end of Park Avenue to the stop sign at the intersection with Fairfield Road at mile point 2.6.

Turn right onto Fairfield Road. Go about 60 yards and turn left onto Willoughby Run Road. Follow Willoughby Run Road as it twists past a few houses and out into the open. As you pass fields on your right, Willoughby Run can be seen to your left through the line of cedar trees and down the slope.

Continue to mile point 3.5 where Willoughby Run Road bends sharply to the right, heading away from the creek and through a modern farm. Use extreme caution passing between the private farm buildings. Continue to the Stop sign at the intersection with Blackhorse Tavern Road at mile point 3.9.

☛ **STOP:** at the Stop sign intersection with Blackhorse Tavern Road. Pull off partway onto the grass shoulder to the right and engage your hazard flashers. Remain inside your vehicle.

☛ **LOOK:** left and right up and down Blackhorse Tavern Road, making note in particular of the open landscape to your right.

Quick Summary: After much tangential effort, you have finally returned to the historic Blackhorse Tavern Road just as McLaws' Division did in 1863.

McLaws' countermarch brought him all the way back to Herr Ridge. The column then followed Old Mill Road before turning off to the right (south) along another long-lost farm lane or fence line. The modern street you used to get here, Park Avenue, runs parallel to that 1863 farm lane, which some historians associate with the nearby Adam Butt Farm.

McLaws

The point where Blackhorse Tavern Road crossed Breams' Hill at Driving Tour Stop #6 – the point where the march came to a halt and this whole countermarch began – is only half a mile to the right of where you are now parked. It would have taken the leading edge of McLaws' Division no more than 15 minutes to cover that distance – but the whole division as it followed would have been under Union observation from Little Round Top the entire time, followed by their artillery battalion, then by Hood's Division and by his artillery battalion.

Instead, the countermarch from Bream's Hill to this spot covered at least four miles, which at a standard marching pace would have consumed at least 90 minutes.

The spot where you are now parked is not on the historic part of Willoughby Run Road. The road you drove along bent to the right, turning away from Willoughby Run and passing through a modern farm. The historic farm lane in 1863 followed the creek more closely. The historic lane merged into Blackhorse Tavern Road about 800 yards to your left. (You will be heading past that spot next.)

More Detail: The Peach Orchard was still a mile away.

Some historians, including respected National Park Service rangers, believe the countermarch did not return to Herr Ridge but simply followed the Fairfield Road east toward Gettysburg until it reached Willoughby Run. This would have been simpler and quicker than the route you just followed. And of course those historians could be right – but the authors do not think so. A countermarch up the Fairfield Road would have brought McLaws over an even higher section of Bream's Hill than the ridge at Driving Tour Stop #6. Today the crest of Bream's Hill on Fairfield Road is bordered by a deep highway cut and by a line of trees that shelter it from observation. Those features would not have existed in 1863.

Consider what has transpired in the time taken to execute Longstreet's initial march from Herr Ridge to Bream's Hill and then the countermarch to this point.

No one can be certain of the precise timing of when anything happened on July 2nd or at any moment during this or pretty much any other battle. Watches were not routinely synchronized and standard time zones, made necessary by the development of railroad schedules, had not yet been created. But what is certain is that hours have passed and that during those hours the Union forces did not remain static. It was during these hours that Major Gen. Daniel Sickles would take it upon himself to redeploy his brigades forward to occupy that one piece of terrain across which Lee had swept his finger on the map that morning: The Peach Orchard.

Driving Tour Stop #8:
Warfield Ridge War Dept. Observation Tower

- **ON THIS LEG OF THE DRIVE:** Your drive will bring you across Willoughby Run, after which the Blackhorse Tavern Road gradually bends right to head due south. (In 1863 this part would have still been on Willoughby Run Road.) As you approach the historic Millerstown Road, the open farm fields on your left are part of the Eisenhower National Historic Site (NHS). This was Pitzer land in 1863 and site of the Pitzer Schoolhouse. It was acquired in the 1950s by Gen. Eisenhower's friend, the oilman W. Alton Jones, who willed it to the National Park Service upon his accidental death in 1962. It served then and today as a buffer for the Eisenhower home farm.

  As you reach the Millerstown Road, you will see a white house across the street. This was the site of the 20th Century replacement schoolhouse and today is a private home, the former residence of Brig. Gen. John D. Eisenhower right next to his father's home farm.

  The last leg of your drive up the Millerstown Road retraces the final approach of McLaws' Division onto Warfield Ridge, providing the general with his long-awaited first glance at the battlefield and the Peach Orchard. It would be an unhappy moment.

- **GO:** Turn left onto Blackhorse Tavern Road. Pass over the bridge crossing Willoughby Run at mile point 0.4. Follow Blackhorse Tavern Road until you reach the Stop sign at Millerstown Road at mile point 0.8.

  Turn left onto Millerstown Road and drive past the white-fence driveway entrance to Eisenhower Farm at mile point 1.1. (Note: Private vehicles are not permitted to drive into Eisenhower NHS.) Continue up the slope, approaching Warfield Ridge from the rear. At the stop Sign, turn right onto West Confederate Avenue at mile point 1.4. Proceed to the War Dept. Observation Tower on Warfield Ridge at mile point 1.5. Park in the lot, leave your vehicle and walk to the pair of large black 20 lb. Parrott Rifles positioned near the base of the tower. (This is the Ashland Artillery of Virginia commanded by Capt. Pichegru Woolfolk.) On the left side of the left gun you will see a notch in the stone wall leading to a short path.
- **STOP:** follow the path trail about 10 yards through the woods to the edge of the field.
- **LOOK:** east across the fields toward the Peach Orchard and the Emmitsburg Road 550 yards in front of you.

<u>Quick Summary:</u> You are standing on Warfield Ridge near the point at which McLaws would discover Union brigades deployed in the Peach Orchard in front of you and north to your front-left along the Emmitsburg Road. The burly Georgian's arrival was promptly greeted by artillery fire.

The predicament was perfectly captured by the words of Confederate Col. Benjamin Humphreys, commander of the 21st Mississippi infantry. He said: "I understood that our march to the Peach Orchard was to bring us onto the enemy's flank. But when we got there, it was unmistakably anything but the flank. It was a

compact, formidable line of frowning artillery and bristling bayonets."

McLaws dispatched couriers yet again to alert Longstreet to yet another crisis. Taking stock, Longstreet quickly adapted his plan. Lee may have been at hand to consult him. It was no longer possible to advance McLaws' Division to the Peach Orchard first. Instead, Longstreet would hold McLaws in position and move Hood's Division into a new line to the right (south). It would now be up to Hood to initiate the attack.

Longstreet

It is also important to note that like the Confederate commanders in 1863, you today also cannot see beyond the Peach Orchard. They had no way to know what the ground looked like beyond that ridge and what sort of Union firepower might be in position or coming into line.

More Detail: Your location on Warfield Ridge is significant in three ways:

1. Command decisions.
2. Artillery combat.
3. Infantry position.

Point One: Command decisions.

You have arrived on the ground at which McLaws would receive his second unwelcome surprise of the day: Union brigades were deployed in the Peach Orchard and north along the Emmitsburg Road to the white-log Klingel Farm and beyond. The Confederate general's arrival was then greeted by artillery fire from 1st New Jersey Battery B. One shell sailed beyond Warfield Ridge and killed a horse in an artillery battery.

You will recall McLaws' first unwelcome surprise of the day. A mile back on Blackhorse Tavern Road hours ago he discovered his march could be observed by Union lookouts on Little Round Top. It required a lengthy countermarch to find a new route and avoid detection. The hill is now easy to see. The soft, wooded peak on the horizon to the right of the Peach Orchard is Big Round Top. Little Round Top is just to the left. Its open plateau and several monuments should be visible.

Now McLaws gets his second surprise – the presence of Sickles' forward Union line less than 600 yards to his front. (In war, surprise is seldom welcome.) A staff officer came up to investigate the sudden delay and McLaws sent him back to find Longstreet. The situation had changed. A message was returned to McLaws insisting there could be nothing more than a Union regiment in front of him and that McLaws should brush it aside. McLaws had this messenger return to Longstreet with the insistence that the corps commander was failing to grasp the problem. This prompted Longstreet to come up to see the situation for himself.

Longstreet quickly applied the adage familiar to any old warrior: "The first casualty of war is the plan."

Plan A, the advance of McLaws' Division to the Peach Orchard to form a battle line on or behind the Millerstown Road, with Hood coming up in support, was a dead letter.

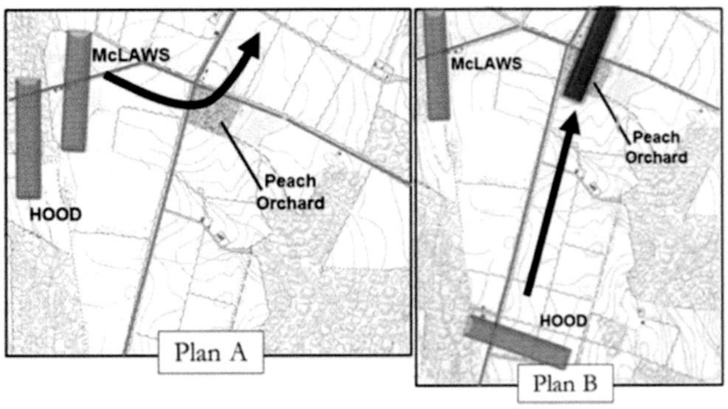

Longstreet on the spot developed Plan B: Hold McLaws in position, move Hood's Division into a new line to the right of McLaws and have Hood start the attack from the southern edge of Warfield Ridge.

Turn around and go back up the path to the big Parrott rifles. Look across the street to survey the open spaces west to the horizon.

The low stone wall a few feet beyond the pavement marks the boundary line between Gettysburg National Military Park and the Eisenhower National Historic Site. Hood's Division, still coming up the Blackhorse Tavern Road, would get new orders to continue straight across the Millerstown Road and through the Biesecker (Eisenhower) Farm. Hood was to deploy south of McLaws along Warfield Ridge on the other side of the Emmitsburg Road.

Move to your right to the low stone wall beyond the base of the tower. Look to your right again toward the Peach Orchard and the Emmitsburg Road.

The overall objective for the assault force on July 2nd was to attack "up the Emmitsburg Road," as Lee would explain in his reports.

Plan A was to have McLaws strike first with Hood in support. Then the divisions of the Third Corps would engage from the left. First the division of Major Gen. Richard Anderson, positioned along Seminary Ridge half a mile in front of you, would advance brigade by brigade. Then, as the "decisive point" on Cemetery Hill was smashed and the Army of the Potomac was reeling, Major Gen. Dorsey Pender would launch his division forward to exploit the breach.

Plan B switched McLaws with Hood. Hood would form his line a half-mile farther south along Warfield Ridge itself, turn north and attack first by marching up the Emmitsburg Road through the Peach Orchard, the Rose Farm and the Wheatfield. Hood's Division was to drive in this new Union flank. McLaws was to follow in support, with Anderson attacking from the left and Pender coiled and ready to pounce.

Second point: Artillery combat.

You are at the position at which the opposing batteries were at their closest during the artillery "gunfight" at the Peach Orchard. (The term "gunfight" is used here to represent close action in which these direct-fire weapons could engage one another. It is not an accepted Civil War-era term.)

The movement of Hood's Division to a new position and the alignment of his four brigades into an attack formation would take time. During this delay, some 30 to 45 minutes, Confederate and Union artillery batteries started banging away at each other, with more batteries on both sides coming up, unlimbering and opening fire.

It is here that Alexander would say: "I hoped to make it short, sharp and decisive, for at close range there was less inequality in our guns and especially in our ammunition, and I thought if ever I could overwhelm and crush them, I would do it now. But they really surprised me, both with the number of guns they developed and how they stuck to them."

At the height of the artillery action, 48 Confederate guns along Warfield Ridge were hammering away at 38 Union guns from the Sherfy Farm through the Peach Orchard to the Devil's Den. The Sherfy Farm is represented by the large red barn on the Emmitsburg Road to your front left with the two-toned Sherfy Farmhouse to the left of the barn. The firing distance from this spot to the Peach Orchard and other points on the Emmitsburg Road was 500 to 600 yards muzzle to muzzle. This action between 86 guns made it the largest close-range artillery "gunfight" of the entire Civil War.

Take a moment to look at the Confederate guns in position today around you, all part of Alexander's battalion.

The massive 20-pounder Parrott rifles near the tower steps behind you represent the Ashland Artillery. The pair of Ordnance rifles to your left front represent the Bedford Artillery. Both of these Virginia batteries were actually held in reserve by Alexander along the Millerstown Road. They did not participate in the initial gunfight.

Farther up West Confederate Avenue, the next pair of Ordnance Rifles represent the "Richmond Artillery" commanded by Capt. William Parker. Parker had a total of four rifles and they were actually positioned right where you are standing by the base of the tower. Look at what a close shot they had at the Peach Orchard from here.

Between Parker's rifles and the Millerstown Road you should be able to spot a pair of green Napoleon cannon

What's a Parrott rifle?

A black cast-iron gun that is rifled with "lans & grooves" inside the barrel to cause the bullet-shaped shell to spin when it is fired. The spin gives it greater range and accuracy. The gun tube has a wrought-iron collar welded around the "breech" at the back for strength. The standard Parrott fires a 10-pound shell. The 20-pounder Parrott fires a 20-pound shell.

What's an Ordnance rifle?

A black, wrought-iron rifled gun, stronger and lighter than the Parrott. It fires a 3-inch diameter, 10-pound projectile.

What's a Napoleon?

A bronze smoothbore gun that fires a 12-pound round shell or solid ball. It was developed in France and is nicknamed for Emperor Napoleon III – not the little Corsican fellow who conquered Europe.

representing the Bath Artillery commanded by Capt. Osmond Taylor, also a Virginia battery. This is the battery that included Sgt. Henry Wentz, who grew up on the John Wentz Farm right across the street from the big red Sherfy Barn on the Emmitsburg Road.

Two more batteries from Louisiana and South Carolina with a total of eight shorter-range howitzers were in position beyond the Millerstown Road. As it happens, those two batteries would suffer the heaviest losses in Alexander's battalion with about 70 men killed and wounded between them.

Semmes

All of the batteries mentioned were part of Alexander's big reserve battalion with a total of 24 guns. McLaws wanted the young officer to hold his entire battalion back because he did not want Alexander's guns to draw Union counter-fire that would send shells into the midst of his troops. Longstreet overruled McLaws and ordered Alexander to bring up and engage his batteries.

<u>Third point: Infantry position.</u>
<u>Turn to face the Peach Orchard.</u>

You are on the ground linking the two brigades which would lead McLaws' attack, Mississippi to your left and South Carolina to your right. Two Georgia brigades were in support behind Warfield Ridge aligned with each of those two front brigades.

To your left, the four regiments of Brig. Gen. William Barksdale's Mississippi Brigade would form a battle line to the Millerstown Road and beyond. As they first arrived, however, the men tried to find the best cover they could in the Pitzer Woods to your left and rear to get clear of the artillery fire. Your exact position was occupied by the 21st Mississippi infantry, which was Barksdale's right-flank regiment. The 21st Mississippi would move from here directly into the Peach Orchard.

Alexander

To your right were the five regiments and one battalion of Brig. Gen. Joseph Kershaw's big South Carolina Brigade. The battalion and two regiments were in the woods in the front and to the right of the Warfield Ridge Tower – the "Tower Woods." Two other regiments were in back of the road in the Biesecker Woods farther south and the last one, the 15th South Carolina, was detached at the other end of the Biesecker Woods near the white Snyder Farmhouse.

Once Hood's attack was fully developed, it would be McLaws' turn to advance.

Kershaw would go first. His men would march to the Emmitsburg Road and across it toward the Rose Farm to the right of the Peach Orchard, eventually going into battle in the Wheatfield in support of Hood's men already engaged there. Once Kershaw was across the road and out of sight, Barksdale would make his attack against the Peach Orchard and the Sherfy Farm.

To your right again, the Georgia Brigade of Brig. Gen. Paul Jones Semmes would advance through and around the Tower Woods to follow Kershaw toward the Wheatfield. Semmes – cousin of the legendary Confederate Admiral Raphael Semmes – would be mortally wounded during this advance. The location of the Georgia State Memorial just beyond the parking lot was selected in tribute to Semmes and his men.

And to your left again, another Georgia Brigade commanded by Brig. Gen. William Wofford would advance in the footprints of Barksdale's men. They would march right to and through the Peach Orchard. But while Barksdale wheeled his regiments to the left up the Emmitsburg Road, Wofford would continue straight east toward Little Round Top.

In the midst of the sudden rearrangement from Plan A to Plan B, it is interesting to note the high level of activity of "Pete" Longstreet himself. He would venture out onto Barksdale's forward skirmish line – an unusually hazardous undertaking for a lieutenant general. Longstreet would ride up with Kershaw to the Emmitsburg Road to ensure that the South Carolina units did exactly as he intended. Longstreet would also be in line with Barksdale as the Mississippians prepared to attack. Finally he would accompany Wofford all the way to the base of Little Round Top. Longstreet was on hand to call off the attack at the end of the day.

Bear in mind Plan B was completely improvised. It is evident Longstreet wanted to keep a sure hand on the controls of McLaws' Division to adapt and execute his plans.

QUICKER TOUR: Skip Driving Tour Stops #9 and #10 and proceed directly to Driving Tour Stop #11 Texas Memorial on Warfield Ridge.

Leaving the War Dept. Observation Tower parking lot, continue driving south on West Confederate Avenue past the Snyder House on your left and the Arkansas State Memorial on your right. You will reach the Stop Sign at the intersection with Emmitsburg Road. Carefully cross the Emmitsburg Road to remain on Confederate Avenue (now called South Confederate) until you reach the Texas Memorial on your right. Pass the fence post and pull off to the side of the road by the War Dept. sign for Benning's Brigade. Engage your hazard flashers and exit your vehicle. Pick up the tour instructions for Driving Tour Stop #11.

Driving Tour Stop #9: South End Tower Woods

- **ON THIS LEG OF THE DRIVE:** Your short drive will bring you across the rear of the position along which McLaws deployed half of Kershaw's brigade. Before you reach the Georgia Memorial you will see a pair of 12-pounder howitzers on your right representing one section of Capt. Henry Carlton's Troup Artillery of Georgia. After the Georgia Memorial stand a pair of Napoleons representing Capt. Basil Manly's North Carolina Artillery Battery A. As you pass the South Carolina State Memorial you will see the second section of the Troup Artillery, a pair of Parrott Rifles.
- **GO:** Pull out of your parking space and continue south on West Confederate Avenue in the opposite direction from the Millerstown Road intersection. (This is a one-way road and you must continue in the same direction you were traveling when you drove into the parking lot.) Pass the Georgia Memorial on your right at mile point 0.1. When you reach the end of the Tower Woods on your left at mile point 0.2, pull over to the right, park and exit your vehicle. Cross the road and carefully climb over the stone wall and walk east in the direction of the Emmitsburg Road. Walk across the edge of the field about 100 yards with the Tower Woods to your left.
- **STOP:** at the corner end of the Tower Woods where the view opens up across the fields to enable you to see the Peach Orchard now 600 yards to your left front.
- **LOOK:** toward the Peach Orchard.

Quick Summary: You have come to Col. Henry Coalter Cabell's artillery battalion line. His four batteries were in position from this spot south to the white Phillip Snyder Farmhouse. They engaged Union batteries in and around the Peach Orchard.

One veteran gunner, Andrew Reese of the Troup Artillery, said, "The firing was the most rapid I have ever witnessed. The ground literally vibrated under the continuous roar."

You are also at the southern edge of the Tower Woods in which half of Brig. Gen. Joseph Kershaw's Brigade was positioned. Two more regiments were in the Biesecker Woods across the street behind you. When Kershaw's Brigade advanced, they moved across the field in front of you toward the Emmitsburg Road.

Union batteries were in and around the Peach Orchard. McLaws wanted Cabell's guns in this position so they would not draw that Union artillery fire into the ranks of his infantry regiments formed to your left and rear.

Kershaw

More Detail: Your position was held by those Confederate batteries which were closest to the Peach Orchard some 600 yards northeast of you.

Cabell's shorter-range bronze guns were Napoleons and howitzers. The four bronze guns were positioned closest to the Peach Orchard. The 10 longer-range iron rifles of the battalion were positioned to their right for the longer shot.

When the South Carolina brigade advanced toward the Emmitsburg Road, Kershaw in his report said Longstreet accompanied him as far as the road, but evidently this did not allay his confusion. Kershaw also expressed dismay that Barksdale's Brigade was still forming its lines back on Warfield Ridge and would not be coming up with him to cover his left flank.

This area can also be seen as the extreme left end of Hood's division. For sure, the Texas Brigade, which led the attack, was quite far to your right at the Snyder Farm and straddling the Emmitsburg Road. But supporting the Texans was Brig. Gen. George "Tige" Anderson's Georgia Brigade positioned to your right rear. When Anderson advanced his brigade, his left regiment was the 9th Georgia infantry. According to the detailed account provided by Capt. George Hillyer of the 9th

Georgia, his men marched directly through some of the guns manned by his fellow Georgians of the Troup Artillery. (Consult the map on page 20).

Anderson's Brigade went in before Kershaw, holding to the left flank of the Texas Brigade all the way to ground below the Devil's Den before veering north to engage Union regiments in the Wheatfield. Kershaw, in turn, would come in on Anderson's left.

Driving Tour Stop #10: Snyder Farm

☛ **ON THIS LEG OF THE DRIVE:** You are driving along the back of the artillery line of Cabell's battalion firing at the Union batteries in the Peach Orchard. You will pass a marker for Capt. Edward McCarthy's Richmond Howitzers battery from Virginia. It features a lone Napoleon cannon. You will also see a marker for Capt. John Fraser's Pulaski Artillery from Georgia.

☛ **GO:** Turn around and return to your vehicle, carefully stepping over the stone wall. Drive south on West Confederate Avenue, passing the Biesecker Woods on your right and following the stone wall on your left. Pass the iron sign identifying Biesecker Woods. When you reach the gap in the stone wall on your left for a dirt trail at mile point 0.2, pull over to the right side of the road and stop. Exit your vehicle, cross the road and walk up the dirt path.

☛ **STOP:** when you have walked about 100 feet on the dirt trail, roughly halfway to the Emmitsburg Road in front of you, turn left and walk out into the field and stop. The white Snyder Farmhouse is now directly behind you.

☛ **LOOK:** north up the Emmitsburg Road toward the Peach Orchard and the intersection with the Millerstown Road.

Quick Summary: You are at the longest range for the guns of Cabell's battalion firing into the Union position in the Peach Orchard. The four rifled cannons of the Pulaski Artillery would have been arrayed in front of you.

You are also in the vicinity of Col. William de Saussure's 15th South Carolina regiment. Separated from the rest of Kershaw's Brigade, they would advance on their own directly east and would fight independently all day.

Turn right to face east.

Thanks to restoration work by the National Park Service, you now have a striking and dramatic view of the Devil's Den and the rocky front face of Little Round Top behind it. Not long ago, woods blocked this view.

If you use your imagination to remove from your view the vehicles, people or monuments that may happen to appear on Devil's Den, you can see how for the Confederate troops in 1863 it was not immediately clear that Devil's Den and Little Round Top were separate features. From this vantage they rather appear to be a single mass of rising stone.

More Detail: This vantage also helps explain why Capt. John C. Fraser's Pulaski Artillery was the hardest hit of the batteries on Cabell's line. On the left edge of the Devil's Den would be the Houck's Ridge position of Capt. James Smith's 4th New York Artillery. Firing from the right front of the Peach Orchard would be the six Parrott rifles of Capt. Judson Clark's 1st New Jersey Battery B, who in his report identified the Confederate battery near the white farmhouse as his first target. Fraser's men would catch hell from two directions. The result was seven killed and the captain himself mortally wounded. The battery's losses in men and horses would be so heavy, their surviving guns would be folded into Manly's North Carolina battery.

Your position also offers you the view of the Gettysburg battlefield first presented to Union officers using the Emmitsburg Road on their ride north from Maryland, including Sickles himself. It would seem that a simple glance to the right here would have revealed Little Round Top to be a crucial defensive position. Yet when Sickles arrived, the battle he would have heard was occurring miles in his front. With eyes front toward the sound of the fighting, the Peach Orchard would have been unmistakable as the critical high point a half-mile ahead on the Emmitsburg Road. And the farther he rode up this road toward the Peach Orchard, the more Little Round Top receded from his view.

Driving Tour Stop #11: Texas Memorial on Warfield Ridge

- **ON THIS LEG OF THE DRIVE:** You are passing along the initial battle line of the Texas Brigade, which initiates the attack to set off the infantry fighting on July 2nd. The brigade included the 3rd Arkansas infantry, which is why you will see the Arkansas State Memorial opposite the Snyder Farmhouse.
- **GO:** Return to your vehicle and continue driving south on West Confederate Avenue past the Snyder House on your left and the Arkansas State Memorial on your right. You will reach the Stop Sign at the intersection with Emmitsburg Road at mile point 0.1. Carefully cross the Emmitsburg Road to remain on Confederate Avenue (now called South Confederate) until you reach the Texas Memorial on your right at mile point 0.2. Pass the roadway gate post and pull off on the right the side of the road by the War Dept. placard for Benning's Georgia Brigade. Engage your hazard flashers and exit your vehicle.
- **STOP:** walk back up South Confederate Avenue to Latham's artillery battery on the right side of the road.
- **LOOK:** north up the Emmitsburg Road toward the Peach Orchard and northeast across the fields toward the Michael Bushman Farm and beyond it toward Little Round Top.

Quick Summary: Longstreet's new plan called for Hood's Division to lead the attack.

At your vantage you are surveying the long Union flank from the Peach Orchard through the Rose Farm and farther right all the way to the Devil's Den to your front right. In front of this flank was a deadly line of skirmishers in the form of the elite, green-uniformed companies of U.S. Sharpshooters.

The notion of moving up the Emmitsburg Road toward the Peach Orchard became untenable.

Union troops and guns on Devil's Den had made their presence known. Hood knew he would have to deal with them first. It also became evident that occupying the heights of Little Round Top would be necessary to any effort to drive in the Union flank in this area.

Soon after the attack commenced, Hood was wounded and out of the fight, deepening the confusion.

More Detail: Moving up the Emmitsburg Road toward the Peach Orchard was a substantial distance but the rolling fields were open and there was a lot of terrain separating the line of attack from Little Round Top to the right.

But at your present position in 1863 you would have seen that this Union flank presented itself in a broader front from the Peach Orchard through the Rose Farm all the way to the Devil's Den. The formidable line of skirmishers held farm lanes behind the Bushman and John Slyder farms.

Any attack route going through the Devil's Den, Houck Ridge and the Wheatfield could be flanked by any troops on Little Round Top. It became obvious the hill, which for the moment was still empty (except for those pesky signalmen), had to be seized as part of this initial assault.

For Hood, the geography of the ground presented a formidable hazard. Sweeping across open fields around the Peach Orchard was one thing. The massive stone formations to his right front were another matter altogether. Hood would impress upon Longstreet the virtue of avoiding these obstacles by moving even farther east and swinging around the back of Big Round Top into the Union rear.

Much is made of this idea in post-war writings and some modern books. For Longstreet, this had to be a non-starter. First, it would separate his two divisions. Second, it would disrupt the alignment of Hood's four brigades, two in front, two in support, which Longstreet felt he needed to ensure the attack had sufficient power. Third, it would stray too far away from Lee's original instructions, which Longstreet was already amending.

There was a silver lining in the darkening prospects of this increasingly skewed attack. If Hood's Division could smash through the Devil's Den and Wheatfield, they would knock the pins out from underneath the Union position in the Peach Orchard. The brigades in the orchard and along the Emmitsburg Road would have to collapse back without a rifle shot. McLaws could march his division forward unimpeded. Plan A would be resurrected.

All such silver-lining optimism aside, consider for a moment how the "decisive point" of the battle – Cemetery Hill – was getting farther and farther away with ever more terrain and ever more Union troops in the way.

Bear in mind something else. Somewhere down this sloping field between your position and the farm lanes behind the Bushman house, Hood would be crippled by an exploding artillery shell. He would be deprived of the full use of his left shoulder for the rest of his life – and Longstreet would be deprived of his best combat commander for the rest of the battle.

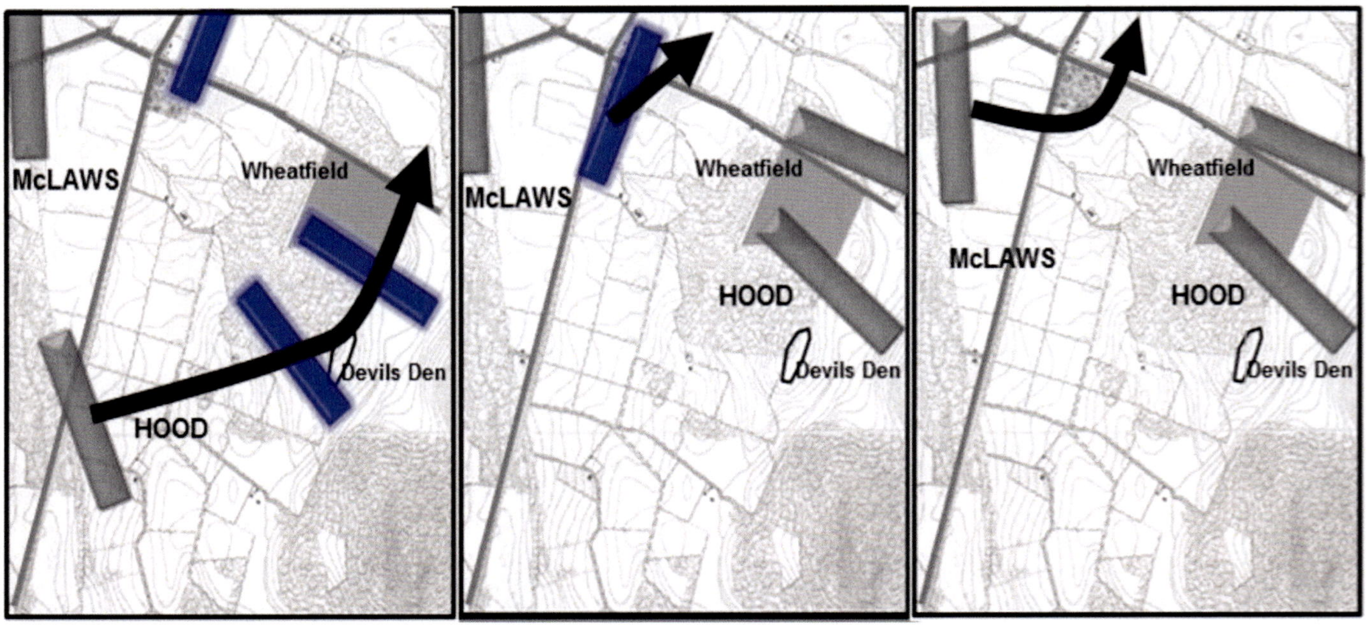

If Hood could smash through and destroy the Union positions to his right in and behind the Devil's Den and Wheatfield, it would knock the pins out from under the Union position in the Peach Orchard, forcing those troops to withdraw. This would enable McLaws to resume the original "Plan A" to form on the Millerstown Road by the Peach Orchard and to attack north toward Cemetery Hill.

QUICKER TOUR: Skip Driving Tour Stops #12 and #13 and proceed directly to Driving Tour Stop #14 Peach Orchard Salient Wayside Sign.

Return to your vehicle, drive up to the Picnic Area about 100 yards ahead on the right, drive in the follow the driveway to make a U-turn. Go back to the Stop Sign on Emmitsburg Road, turn right and drive 0.6 mile to the Peach Orchard. Turn right onto Birney Avenue, the small lane that borders the Peach Orchard on the south and east. Drive down to the turn, follow the turn left, drive up and park anywhere on the right side with your car facing north toward the Wheatfield Road. There is a gravel strip so you can pull partly off of the road. Park there and pick up the tour instructions for Driving Tour Stop #14.

(Never park on the Wheatfield Road itself under any circumstances. It is illegal and hazardous.)

Driving Tour Stop #12: Brook Avenue Sign/The Wheatfield

☛**ON THIS LEG OF THE DRIVE:** You will pass along the line of the Law's Alabama Brigade, which advanced on the right of the Texas Brigade. As South Confederate Avenue passes over Big Round Top, you will be following their advance toward Little Round Top.

As you come down the slope from Big Round Top and approach the south side of Little Round Top, you will come upon the line formed by the Union brigade of Col. Strong Vincent, newly arrived. His regiments would open fire into the face of the Alabama men, setting off one of the most famous fights of the battle.

Passing over Little Round Top, then left on Wheatfield Road and left again into the Wheatfield itself, you will be seeing one of the bloodiest landscapes of the Civil War, especially the Wheatfield. The Wheatfield fighting is usually described to include the advance of Col. John Brook's brigade up on the wooded ridge west of the Wheatfield. You will pass along this ridge where Semmes' Georgia Brigade would confront Brook's line.

☛**GO:** Return to your vehicle and continue forward (south) on South Confederate Avenue heading away from the Emmitsburg Road. Pass the Alabama State Memorial and the Soldiers-&-Sailors-of-the-Confederacy Memorial on your right, and then finally the position of Capt. James Reilly's Rowan Artillery battery from North Carolina at mile point 0.3. Continue on South Confederate Avenue as it bends sharply east.

Pass the Timbers Farm/Cavalry Battle wayside markers on your left at mile point 0.7. Continue northeast up the slope of Big Round Top and pass the parking area at mile point 1.1. As you come down the slope, cross Warren Avenue at mile point 1.4 and drive up the side of Little Round Top, passing through the parking area at mile point 1.6. Continue down to the Stop Sign at Wheatfield Road at mile point 1.8.

Turn left onto Wheatfield Road. Pass the John Weikert Farm on your right and Crawford Avenue on your left. Cross the narrow bridge over Plum Run. Turn left onto Ayres Avenue at mile point 2.1 to enter the Wheatfield. Follow Ayres Avenue into the woods as it curves past the U.S. Regulars monuments, continue past the 5th New Hampshire monument on your left and come to the Stop Sign at Sickles Avenue at mile point 2.5.

Cross Sickles Avenue and continue straight. The road changes names to Cross Avenue and then Brook Avenue as you follow it along the edge of the woods. Pass the Union monuments on your right and the wayside markers for the Rose Farm on your left at mile point 2.9. As the road slopes down, an iron sign for "Brook Avenue" appears ahead on the left side of the road at mile point 3.0. The Peach Orchard comes into view up the slopes to your left front as the woods thin out.

☛**STOP:** Pull over to the left side of the road on the solid shoulder before reaching the Brook Avenue sign. Leave room for other vehicles to pass. Engage your hazard flashers. Remain inside your vehicle.

☛**LOOK:** up the slopes to your left front toward the Peach Orchard.

Quick Summary: No other point in your driving tour makes it as clear as this spot that the Peach Orchard is high ground. Nowhere else in the national park is the view of the Peach Orchard elevation this dramatic.

The ridge you have driven along for the last 150 yards was the position held by Union troops of 24-year-old Col. John Rutter Brooke's brigade. Brig. Gen. Paul Semmes' Georgia Brigade would strike them here, marching across the Rose Farm to your left. It is in this area that Semmes himself would be mortally wounded.

To your front across the open gap is the rocky, wooded position known as the Stony Hill. That is the position where Kershaw would lead two regiments in support of Anderson's Brigade, which was fighting in the Wheatfield to your right.

Driving Tour Stop #13:
The Loop/Stony Hill

- **ON THIS LEG OF THE DRIVE:** You will pass through the bloody Wheatfield battlefield along the front of the line where Anderson's Georgians would attack from the woods to your right and push the regiments of Col. Regis de Trobriand's Union brigade north, only to be driven back in turn by divisions of the Union Fifth Corps and Second Corps. This was hard fighting because if a Confederate unit were to break through this field, it could move directly toward Cemetery Ridge from the south and in the process cut off all of the Union forces at the Peach Orchard and up along the Emmitsburg Road.
- **GO:** Drive forward on Brook Avenue into the Wheatfield to the Stop Sign at Sickles Avenue at mile point 0.2. Turn left onto Sickles Avenue and drive into the woods, passing the famous Irish Brigade New York monument on your left at mile point 0.3. Follow the road carefully as it bends sharply left and then bends again to the right through the Stony Hill section. This part of the modern tour road is called, appropriately enough, "The Loop." As you complete the sharp bend and pass the iron sign for The Loop on your left at mile point 0.4, pull over to the right side of the road by the pair of Union monuments.
- **STOP:** park. Leave your vehicle, cross the road and walk over to the left side of the monument to the 116th Pennsylvania infantry, which has a granite sculpture that depicts a fallen soldier lying behind a stone wall.
- **LOOK:** Across the fields in front of you up the slope toward the Peach Orchard.

Quick Summary: This spot at the Stony Hill provides yet another vantage illustrating that from the Union perspective, the Peach Orchard is the high ground.

In the positions initially taken by the Union Third Corps, the 3rd Michigan's long skirmish line proceeded from the point of woods to your left all the way up to the near corner of the Peach Orchard, which is where their monument is located. It is evident from this perspective that no single regiment could hold such a line for long.

The ground before you slopes up steadily toward the Peach Orchard. This was the scene of the disastrous change of direction taken by some of the South Carolina men who were advancing against the Union batteries on the Wheatfield Road to your right. Confused orders caused them to halt, change direction and continue their march directly toward your position. In the process of making this erroneous turn, they exposed their flank to devastating artillery fire.

More Detail: This is where Kershaw led his two center regiments, the 3rd and the 7th South Carolina, across the Rose Farm, which is beyond the woods to your front left.

Your drive up from the Wheatfield through the woods behind you brought you past the most famous Irish Brigade monument with its Celtic Cross formed of green marble. However, the monuments for the 116th Pennsylvania next to you and for the 28th Massachusetts across the road behind you also memorialize units of the Irish Brigade. The Celtic Cross monument commemorates only the New York regiments. In fact, the 28th Massachusetts had as many men in its ranks as the three New York regiments combined.

Your position is slightly to the right and rear of the curved line of the 118th Pennsylvania, the Philadelphia unit raised by the famous Corn Exchange, which at 225 men was the largest of the regiments in Col. William Tilton's undersized brigade. The 118th faced both south and west as they held the woods of Stony Hill between your position and the Rose Farm beyond. Kershaw's two regiments would slam into their line through these woods. Having made contact, Tilton would pull his men back. Kershaw would follow along the road to your right and press the men of the Union Fifth Corps back toward Wheatfield Road.

Driving Tour Stop #14:
Peach Orchard Salient Wayside Sign, Northeast Corner

- **ON THIS LEG OF THE DRIVE:** Turning left onto the Wheatfield Road, you will pass along the first position of Capt. John Bigelow's legendary 9th Massachusetts Artillery, demarked by a pair of Napoleon guns, and of the 5th Massachusetts battery, demarked by a pair of Ordnance Rifles. These were the last Union batteries coming into position for the artillery gunfight, engaging with Confederate batteries south and west along Warfield Ridge. The Trostle Barn visible to your front before you turn and to your right after you turn will mark the location of Sickles' headquarters.
- **GO:** Drive forward on Sickles Avenue to the Stop sign at Wheatfield Road at mile point 0.1. Turn left onto Wheatfield Road and head west toward the Peach Orchard until you pass the orchard on your left. You will reach a Stop sign on Emmitsburg Road at mile point 0.35. Turn left onto the Emmitsburg Road, go 150 yards and turn left again onto Birney Avenue, the small lane that borders the Peach Orchard on the south and east. Drive down to the turn, follow the turn left, drive up and park anywhere on the right side with your car facing north toward the Wheatfield Road. There is a gravel strip so you can pull partly off of the road.

(Never park on the Wheatfield Road itself under any circumstances. It is illegal and it is hazardous because this is a through road with two-way traffic and it crests at a blind hill. This is why the official NPS Auto Tour Stop 10 for the Peach Orchard is actually back along Sickles Avenue – our next stop.)

Walk up the asphalt steps with handrails at the northeast corner of the Peach Orchard to ground level.

- **STOP:** at the top of the asphalt steps.
- **LOOK:** toward the center of the Peach Orchard and toward the wayside sign directly in front of you.

Quick Summary: You can see here again how the Peach Orchard is prominent high ground with open, rolling fields that drop off in every direction, especially behind you. Here you can also see how the orchard rows provide space for men and wagons. The trees in 1863 were high enough to offer concealment.

To your immediate right is a thick, plain obelisk monument commemorating the 141st Pennsylvania Infantry regiment. The men of that regiment sacrificed themselves in a collapsing position to purchase a few minutes so the artillery batteries in the Peach Orchard and on the road could limber up and get out.

The wayside sign in front of you features a painting by a correspondent who witnessed the battle. It also has a portrait of Kershaw.

Madill

More Detail: Turning around, you will see the minie-ball shaped monument to the 7th New Jersey Infantry about 200 yards down the slope. In line behind it is the famous Trostle barn with the prominent cannonball hole on its brick face. Half a mile beyond it, the white dome of the Pennsylvania Memorial guides your eye to Cemetery Ridge and the Union battle line.

Look left up the Emmitsburg Road north toward the town.

You can see how the road is generally at level with the Peach Orchard as it runs toward the white log house of the Klingel Farm while the landscape to the right (east) of the road clearly slopes down. This feature is sometimes referred to as the Emmitsburg Road Ridge.

Look farther left (west) across the Emmitsburg Road to Warfield Ridge in the distance.

That was the left end of the Confederate artillery position.

Now look back at the Trostle Farm.

The Trostle Farm was the origin point from which the Union artillery batteries came into action, charging up the slope toward the Emmitsburg Road.

The two views taken from this position show how the Confederate gunners to your far left could not see over the Emmitsburg Road, could not see the barn and so could not see what might be coming up the slope. Likewise, the Union gunners and brigades racing up that slope toward the front could not see what was waiting for them beyond the Emmitsburg Road.

Turn back around to face the Peach Orchard, then look left to the southeastern corner.

As the artillery fight developed, the corner of the orchard 150 yards to your left was occupied by six guns of 1st New Jersey Battery B. These were the first Union guns to go into action here. Four Napoleons of the 15th New York Battery came in later to the right of the New Jersey guns. These 10 guns would exchange fire with Confederate guns in the positions you have visited along Warfield Ridge.

Beyond the Peach Orchard and across the farm fields you can see Warfield Ridge half a mile south of you. The first Confederate infantry units to advance on July 2nd were those four regiments of the Texas Brigade moving across that distant ground and promptly coming under lethal flanking fire that probably brought down Hood. While the wounding of the Texas general is often attributed to the fire from a Union battery above Devil's Den, the guns of 1st New Jersey Battery B had a superior angle to shoot at Hood's battle line. It can never be known but it is more plausible that a New Jersey shell took the general out of action, condemning his division to confusion and disarray for the remainder of the fighting on July 2nd.

Still facing the Peach Orchard, now look right toward the northwestern corner next to the crucial intersection of the Wheatfield Road and Emmitsburg Road.

It may be helpful to step to your right or to take a knee but you should be able to spot the green Napoleon guns straddling the monument for 1st New York Battery G, commanded by Capt. Nelson Ames. It is from that direction that the attack of the 21st Mississippi came into the Peach Orchard itself.

Across the Wheatfield Road is the monument to the 68th Pennsylvania, known as the "Scott Legion." Their line was on the Emmitsburg Road through the farmyard of the John Wentz Farm across the street and beyond. (The outline of the Wentz foundation is still there.) North across the Emmitsburg Road is the large red barn and beyond it the farmhouse of the Sherfy Farm. That was the position of the 114th Pennsylvania, known as "Collis' Zouaves" for their colorful French Algerian uniforms of baggy red trousers, blue jackets and red stocking caps. They covered the right flank of the 68th Pennsylvania and the 68th covered their left.

The Union infantry line from here to the Sherfy House is the line Barksdale's men would hit once their powerful attack began. By that time, Ames' battery had withdrawn to the ridge behind the Trostle Farm. McGilvery's reserve battery, 1st Pennsylvania Battery C & F commanded by Capt. James Thompson, would have come into the Peach Orchard and Wentz Farm in their place – just in time to be overrun by Confederates.

Along the northern edge of the orchard are the twin monuments for Battery C & F, each with a pair of guns facing into the trees. They were two separate batteries until merged in May 1863. After the war their veterans would return to dedicate separate monuments. The Battery F monument is more memorable with a beautiful statue of a rammer facing west into the sunset. Battery C monument behind it has a granite carving of a cannon. On the day of battle, Thompson's guns were actually all over the place, including within the Peach Orchard and on the far side of the Wentz Farm.

The Peach Orchard was overrun by the 21st Mississippi, commanded by Barksdale's senior and most trusted officer, Col. Benjamin Grubb Humphreys. From your perspective, the 17th and the 13th Mississippi were to the right of the 21st Mississippi. As the Mississippi wave struck, the Union men put up a hard fight but the line dissolved, the 68th Pennsylvania giving way first followed by the 2nd New Hampshire. Brig. Gen. Charles Graham, wounded and his horse killed, would hit the ground and be taken prisoner.

B. Humphreys

The recessed segment of the Birney Avenue lane behind you is the final position of the 141st Pennsylvania regiment, commanded by Col. Henry Madill. That unit is commemorated on the monument next to you, which you may agree is ordinary and nondescript. On a typical battlefield tour by car or bus it might seem to be one of those monuments that is easily missed.

It should not be.

As the Mississippi troops came plowing through the trees in front of you, these Pennsylvania men opened fire and brought their advance to a cold stop. This won precious minutes for the artillery crews in the orchard and on the Wheatfield Road to get their guns out. Those minutes were purchased in blood as the 141st Pennsylvania lost 149 of 209 men.

The landscape around you, from Little Round Top and the Devil's Den through the Wheatfield and across the fields below you to the Peach Orchard itself, is the amphitheater of the bloodiest four-hours fighting of the American Civil War. And on this gruesome day, no Third Corps regiment – none – suffers a heavier percentage of losses (71%) than the 141st Pennsylvania.

As the Mississippi Brigade swept the Peach Orchard, three regiments wheeled to your right to advance their battle line northward. Perhaps Col. Benjamin Humphreys could even spot the fiery Barksdale waving his hat frantically and urging his men forward. The aroused Mississippians would flank and turn the Union regiments out of the Sherfy Farm on the other side of the Emmitsburg Road and then strike the next line of Union regiments forming by the white-log Klingel Farm.

Col. Humphreys knew the plan had been for him to turn his regiment as well and to join in this second phase of Barksdale's ambitious charge. But Humphreys was a combat veteran and now the battlefield conditions compelled a change in tactics.

<u>Turn back around and look down the slope toward the Trostle Barn.</u>

Picture the chaotic scene of retreating troops and batteries limbering with artillery officers racing back and forth across the field barking orders. Visualize the commotion. Imagine what decision you might have made.

No competent commander would pivot his regiment to the north when doing so would leave this congregation of Union troops and guns in his rear, no matter how disorganized the Billy Yanks might appear at the moment.

Col. Humphreys held to the line of attack that had brought him through the Peach Orchard. He did not turn. He advanced his regiment straight ahead down the Wheatfield Road and to the northeast down the slope toward the Trostle Farm. He had the numbers and he had the momentum to drive those Union forces and to sweep the battlefield. The valiant Barksdale, who as a former congressman was one of Mississippi's great "fire-eater" radicals, would simply have to make do with his other three regiments as he turned left to roll up the Union line on Emmitsburg Road and win the war all by himself.

<u>Turn back around to your original position facing into the orchard.</u>

Take a look at the wayside sign entitled "The Peach Orchard Salient." It has a handy image of Kershaw and it explains the movement of Sickles' corps followed by the attack of Kershaw's Brigade. The map in the right corner illustrates the Union lines in blue and Confederate line in red with respect to the Peach Orchard and the adjacent roads.

The painting by Edwin Forbes (uncredited) depicts the dense Peach Orchard as Sickles, mounted in the center, rides boldly forward with his staff to inspect his lines. The white headquarters flag behind him bears the red diamond of the 1st Division, Third Corps. It is not

known which regiment Forbes might have observed lying prone in two lines so far out toward the Rose Farm, perhaps the 3rd Michigan, but the artillery depicted shows the position of 1st New Jersey Battery B. As an artist, Forbes also faithfully depicts what is missing in so many artworks and movies: The thick crowd of horses that would have been standing behind every battery in action.

The authors have spent countless hours on the battlefield admiring and studying wayside signs like this one, especially when they were preparing for the licensed guide exam, which they took together in December 2002. We regard the waysides with affection and we are glad they are here. That said, it is always great fun to find errors. Make note of them and on your next visit share them with your friends – it makes you look smart.

This sign has three mistakes:

1) The Peach Orchard was owned by the Rev. Joseph Sherfy, the first Adams County farmer to be officially listed as a fruit grower, which today is the county's main industry. He would have been amused to see the federal government thought his son John owned the orchard.
2) Barksdale's Brigade shattered the Union lines in the Peach Orchard, then wheeled north to continue their attack. Brig. Gen. William Wofford's Brigade followed, swept through the Peach Orchard unopposed and continued on a separate line east. It is incorrect to imply Wofford's men participated in Barksdale's attack.
3) The "you-are-here" point shown on the map is most decidedly not where you are. It refers to the National Park Service Auto Tour Stop #10, which is about 150 yards behind you on Sickles Avenue. The wayside sign was probably designed to be installed back there at the official tour stop rather than here in the actual Peach Orchard.

Driving Tour Stop #15:
Sickles Avenue/NPS Auto Tour Stop 10 "Peach Orchard"

- **ON THIS LEG OF THE DRIVE:** Leaving the Peach Orchard and the fighting position occupied by 1st New Jersey Battery B, you will cross the downward-sloping ground to the location of the New Jersey battery's actual monument.
- **GO:** Using extreme caution with regard to traffic coming from your left, turn right onto Wheatfield Road, go 120 yards and turn left onto Sickles Avenue. Drive up to NPS "Auto Tour Stop 10," which provides marked parking spaces on the right side of the road at mile point 0.13.
- **STOP:** park in the spaces for NPS Auto Tour Stop 10. Exit your vehicle and cross the road to the artillery monument marked by a pair of Parrott rifled cannon.
- **LOOK:** toward the Peach Orchard.

Quick Summary: The monument here commemorates Capt. Judson Clark's 1st New Jersey Battery B, the very first Union battery to move up into the Peach Orchard in the afternoon. It would be more heavily engaged in the "gunfight" than any other battery. However, if you look toward the Emmitsburg Road in front or left toward Warfield Ridge, you will realize you cannot see far in either direction. The landscape and the orchards obstruct your view. Wherever Clark's battery might have been located for the severe artillery gunfight and the ensuing battle on July 2nd, it is obvious as you stand here this could not have been the spot.

Clark

When the Mississippi Brigade smashed the Union defenses in the Peach Orchard and Sherfy Farm, Barksdale pivoted three of his regiments northward to continue the attack across the ground where you are standing. This attack rolled from the Wheatfield Road on your left across Excelsior Field in front of you toward United States Avenue on your right. Union regiments would make their final stand on United States Avenue and then on a second line beyond it. This second line, formed in rear of the white-log Klingel Farm, was anchored on the big 120th New York Regiment, the "Ulster Guard." Their monument is the stone column that you can see out in the field by itself about 350 yards north of you on the last leg of Sickles Avenue.

More Detail: Read the inscriptions for the monument to Clark's battery. They will tell you that these men fired 1,300 rounds on July 2nd. The actual number was 1,342, the most rounds fired by any Union battery on any single day in the entire Civil War.

Look west toward the Emmitsburg Road. You cannot see the road itself nor anything beyond it. A battery in this position could not fire toward Warfield Ridge at the point where McLaws first arrived, which according to McLaws is exactly what occurred.

Look south across the Wheatfield Road along the left edge of the Peach Orchard. You cannot see Warfield Ridge, the Snyder Farm or the position occupied by Cabell's Confederate artillery battalion south of the Tower Woods. The elevation of ground and the orchard itself obstruct your view in that direction.

Conclusion? Wherever Clark's battery might have been located for the artillery gunfight, this could not have been the spot. So why then is the monument positioned here?

There is some thought by historians with expertise in artillery that this was the location where Clark's battery first deployed "in line," with guns still hooked onto the limber wagons, preparing to go into action but not yet engaged.

Reports contained in the New Jersey Civil War archives concerning the dedication of monuments for that state suggest another reason.

Turn around and look eastward across the road.

Prominent in the field is the large, polished granite minie-ball monument to the 7th New Jersey infantry. The regiment was engaged all around this ground but the monument is placed there to honor their commander, Col. Louis Francine, who was mortally wounded at that spot. Francine was the highest-ranking New Jersey soldier to die at Gettysburg and it was considered a priority in 1888 to commemorate that spot with his regiment's monument. Statements recorded at the dedication indicate that the 7th New Jersey had been assigned "to protect Clark's Battery." This

stretches the facts a bit in the context of 1863 but in the context of 1888 it was used ceremonially to enhance the regiment's role in the battle. As a result, it became important for the battery which they were "protecting" to have its monument placed in the vicinity. Infantry monuments took precedence over artillery units since regiments would leave home with upwards of 1,000 men while a battery comprised just one company. It seems the priority in 1888 was for the 1st New Jersey Battery B monument to complement the placement of the 7th New Jersey infantry monument.

Some battlefield skeptics can take this as further evidence that many monuments at Gettysburg are in the "wrong" place. There is a better way to look at it. With a few notable (and sometimes amusing) exceptions, most of the monuments at Gettysburg are in the "right" place because they are located where the Civil War veterans who fought here wanted them to be placed. Classic examples are the scores of monuments which veterans placed along "avenues" so that the visitors could see them. The designation of "avenue" means it is a park road that did not exist at the time of the battle.

Battles are flowing, fluid events. So too the positions of units involved in those battles are flowing and fluid. No single stone edifice can replicate this reality.

Driving Tour Stop #16: Trostle Farm/9th Massachusetts Battery Position

- **ON THIS LEG OF THE DRIVE:** You will follow the advance of the first three Mississippi regiments across the ground held by Union Col. William Brewster's "Excelsior Brigade." The domed Excelsior monument with five columns stands across the road from Driving Tour Stop #15. Markers for the individual Excelsior regiments are along the fence line to your left.

  As you turn right on United States Avenue, you will be following the advance of the 21st Mississippi on your right. The 8th Alabama infantry, oddly enough another Confederate regiment that happened to be separated from its brigade, would be advancing in the same direction on this road across the ground to your left.
- **GO:** Drive forward on Sickles Avenue away from the Peach Orchard to the Stop Sign at United States Avenue at mile point 0.1. Turn right into United States Avenue and drive down the slope to the Trostle Farm. Pass the vertical-cannon marker for Gen. Daniel Sickles' headquarters position at mile point 0.3.
- **STOP:** park in the marked spaces across from the Trostle Farm. Leave your vehicle and walk up the road about 40 yards until you are opposite the monument and pair of Napoleon cannon positioned at a remnant of the historic stone wall.
- **LOOK:** Turn and look up the sloping fields toward the southwest in the general direction of the Peach Orchard and your most recent Driving Tour Stop, neither of which you can actually see because the rising field limits your view of the horizon.

<u>Quick Summary:</u> You are on the ground of the climactic last stand of Capt. John Bigelow's 9th Massachusetts Battery. The Napoleon guns behind you represent the battery's second position after retreating from the Wheatfield Road across the fields in front of you. They withdrew by using their "prolonge" rope to pull back the guns as they recoiled from being fired.

Bigelow's battery was then ordered to sacrifice itself in order to stunt the advance of the Mississippi infantry.

Look up the slope. You can see that the distance to the horizon is quite limited from this vantage. You are at the bottom of a small depression that forms a half-bowl. This short horizon enabled Bigelow's gunners to keep the Confederates at bay by using canister and case shot with short fuses and by firing solid cannonballs low to make them skip across the ground. Despite such heroics, the battery was overrun with most of the officers and a third the men killed or wounded.

McGilvery had ordered Bigelow to sacrifice his battery to give McGilvery time to assemble a larger line of guns on the ridge to your left on the other side of Plum Run.

Bigelow

<u>More Detail:</u> Bigelow's 9th Massachusetts Battery first held a position on Wheatfield Road. You passed their monument and Napoleon guns between Driving Tour Stop #14 and Stop #15. Untested in battle before July 2nd, the gun crews of Bigelow's battery exhibited exceptional ability, courage and devotion to duty, especially the junior officers, all of whom were killed or wounded. Bigelow himself was badly wounded in the gut and had to be led off the field on his horse by his bugler – an act of

heroism that would earn that bugler, Charles W. Reed, the Medal of Honor.

Aided by Alabama men firing and charging down the Trostle Lane to your right rear, the Mississippi men did manage to overrun Bigelow's guns, capturing all but one of them – but only for a short while. The Mississippi men lacked the equipment to work the guns or the horses to move them. A famous photo on the wayside marker by the Trostle barn shows there were indeed plenty of horses all around here. Their corpses littered the ground.

QUICKER TOUR: Skip Driving Tour Stops #17 and #18 and proceed directly to final Driving Tour Stop #19 "The Upper Angle."

Follow United States Avenue east to the Stop Sign. Turn left on Hancock Avenue and follow it past the Pennsylvania Memorial on your right, passing the intersection with Pleasanton Avenue. Continue on Hancock Avenue past the cluster of Union monuments. Pass the U.S. Regulars Monument, a large obelisk on a square base. Pass the Copse of Trees and the Highwater Mark Memorial on your left. Pass the first marked parking spaces for NPS Auto Tour Stop #15 "The Angle." Continue to the part of the road where the wall bends north and continues close on your left, parallel with the road. Park there and pick up the narrative for Driving Tour Stop #19.

Driving Tour Stop #17:
NPS Auto Tour Stop 11 Plum Run/McGilvery's Artillery Line

- **ON THIS LEG OF THE DRIVE:** You will cross the farthest point of the advance of the Col. Humphrey's Mississippi troops. As you go over Plum Run you will cross over onto the Union line on the small ridge on the other side. The line was held by nothing more than cannon, albeit a lot of them.
- **GO:** Drive past the Trostle Farm house, crossing Plum Run. Continue eastward up United States Avenue until you reach the parking spaces for the wayside sign that represents NPS Auto Tour Stop #11 at mile point 0.1.
- **STOP:** park in the marked spaces and exit your vehicle. Walk back on the road toward the Trostle Farm for about 25 yards. The elevated ground at your left forms a soft ridge above the deeply recessed Plum Run concealed beneath the line of bushes in front of you. Turn left, leave the road and walk out onto that ridge. Depending on the undergrowth and the work of maintenance crews to keep this ground mowed, you may be able to walk all the way over to the snake-rail fence along the restored farm lane. Some well-placed flat rocks might help open the way and give you a clear place to stand.
- **LOOK:** Look back toward the Trostle Farm and across the fields sloping up to the west toward the Peach Orchard.

Quick Summary: You are on part of the ridge on which McGilvery formed what has come to be known as the "Plum Run Line" of guns. When these guns let loose with a sheet of flame, the Mississippi attack that started a mile back on Warfield Ridge would be stopped but cold. It was the end of the Confederate advance in this sector of the battlefield on July 2nd.

McGilvery's guns were deadly in their own right but they also became a soldier's trick. It was customary military procedure to assign infantry to protect artillery. If the Union had put such a line of guns from five or six batteries across this ridge, it meant there had to be a whole brigade of infantry in the woods directly behind them.

McGilvery

There was of course no one behind McGilvery at all. The woods were for the moment empty. McGilvery's intuition, skill and willingness to gamble were the only barrier against a disastrous Confederate breakthrough to Cemetery Ridge and the Taneytown Road. A breakthrough there might have split the Union army in half.

More Detail: McGilvery's "Plum Run Line" included the position of 5th United States Battery I, which probably straddled the farm lane next to you (with the fences removed.) This battery of U.S. Army regulars, commanded by a newly minted West Point graduate, held the left end of McGilvery's line.

McGilvery's line is often interpreted as forming to the right (north) from your position across United States Avenue (which did not exist) and into the fields beyond.

This might be an oversimplification. One report from a battery commander, Lt. Edwin Dow, said his guns were positioned closer to the woods behind you. Also, it has never been entirely clear which batteries or sections were in this line since McGilvery picked off guns as they were retreating past the Trostle Farm, ordering them into position despite the fact some of those guns were not under his command.

Willard

Whatever the truth about such details, the fact was that a heavy force of iron and bronze muzzles formed here in support of one another and they were all aimed the same direction – straight at the Mississippians by the Trostle Farm and on the fields next to it. And this whole gun line was under the command of a single man, McGilvery, a 39-year-old sea captain from Searsport, Me., who knew how to handle guns, how to handle men and how to keep cool under fire.

There was no end to the drama. The guns positioned in the farm lane, 5th U.S. Battery I, would be overrun and captured. Some of the men from the 21st Mississippi would get through the muddy thicket and across Plum Run and were able to charge up this slope to capture those guns. They might have been helped by the fire or bayonets of some of Kershaw's South Carolina men coming in from your left through the Trostle Woods across the marshy ground. The U.S. battery's young officer, Lt. Malbone Watson, would be badly wounded in the leg, which had to be amputated.

<u>Turn to your right (north) to look toward United States Avenue.</u>

In the fields and thickets across United States Avenue and behind the Trostle Farm, an entirely separate and much larger fight would unfold. The three original regiments of Barksdale's Mississippi Brigade would shatter the Union line behind the Klingel Farm, then pivot once again to face toward the east in pursuit of retreating Union troops. This was the third and final leg of Barksdale's famous charge as his men formed battle lines and advanced from the Klingel Farm on the Emmitsburg Road eastward toward Cemetery Ridge, which is about 500 yards to your right. The attackers had a clear view of the Michael Frey Farm on the Taneytown Road. If they could reach it, they would split the Union army's defensive line asunder.

"The Harper's Ferry Cowards" stood in their way.

Freshly redeployed, the men of four proud New York regiments – many of whom had been made prisoners in the humiliating surrender of Harper's Ferry the previous September – now had something to prove. Three regiments under Col. George Willard advanced with fresh vigor and full loads of ammunition to confront Barksdale's men, who for all of their glorious success had to be running short of breath and bullets. The two lines collided with ferocity on the ground behind the Trostle Farm leading into the Plum Run thickets across the road from you. As on McGilvery's Plum Run Line, the Mississippi attack over there would be stopped cold. The bombastic Barksdale was shot from his horse and taken prisoner, mortally wounded.

<u>Turn right again to look east toward the woods. United States Avenue is now on your left</u>

Willard dispatched his fourth, smallest regiment, the 39th New York infantry, to a separate task. The immigrant warriors of the so-called "Garibaldi Guards" charged from the woods in front of you and recaptured those Union guns at the point of the bayonet. Their charge was so vigorous, they trampled one of their own officers, Capt. Carlos de la Mesa, after he had fallen with a wound in the foot.

<u>Turn back around to face west toward the Peach Orchard.</u>

As the Mississippi advance in this sector stalled, Alexander brought up his battalion of 24 guns and formed a firing line from the Peach Orchard across Excelsior Field to United States Avenue. They sprayed artillery fire in every direction into the Union forces. The artillery fire tore off most of Sickles' leg, punched a hole in the Trostle barn which we can still see today and killed Col. Willard just after he had completed his counter-charge to halt Barksdale. McGilvery's gunners elevated their barrels to answer this Confederate firing line. An artillery duel between these two lines across 750 yards of sloping ground would continue past dark. The gunners fired through the blackness by taking aim at the spray of sparks from the opposing muzzles.

It might occur to you that such a dramatic event as the loss and recapture of 5th U.S. Battery I ought to be commemorated with a monument. It was. Just not here.

As you return to your car, you will see a tablet-style monument out in the field 100 yards to the north side of United States Avenue. The shape is the same for every government-issue monument for a U.S. Army unit at Gettysburg. This one is for 5th U.S. Battery I. A wedge-shaped marker in front of larger monument was placed by the veterans of the 39th New York as tribute to the heroics of their men who charged forward under fire to recapture that battery. Those two monuments are indeed in the "wrong" place. They may be the best example of how the positioning of monuments can mislead historians and scholars as they interpret the events of the battle.

Driving Tour Stop #18: 1st Minnesota Memorial

- **ON THIS LEG OF THE DRIVE:** You will be arriving on Cemetery Ridge, that critical line. As you drive along Hancock Avenue you will traverse the part of the ridge that had been left open by the advance of Sickles men and then by the movement of Second Corps troops to the Wheatfield. Gaps in the woods to your right show how close this line was to the crucial Taneytown Road.
- **GO:** Drive up United States Avenue, passing the George Weikert Farm on your right. Drive to the Stop Sign at the T-intersection with Hancock Avenue at mile point 0.2. Turn left onto Hancock Avenue and drive past the line of batteries and monuments along your left until you reach the Minnesota Memorial at mile point 0.5. It is a tall rectangular pedestal with a statue of the soldier advancing with his bayonet out front.
- **STOP:** park on the right side of the road in the marked spaces just past the Minnesota Memorial. Exit your vehicle and walk over to the right (north) side of the Minnesota Memorial, stepping onto the low earth berm that runs along the road.
- **LOOK:** Survey the battlefield landscape in front of you, looking toward the Peach Orchard now a mile distant to your southwest.

Quick Summary: Driving Tour Stop #8 by the War Dept. Tower on Warfield Ridge gave you a panoramic view of this epic battlefield from the Confederate perspective. Now you have a panoramic mirror view from the Union perspective.

The Peach Orchard is easily seen to your front left. From this vantage, the orchard is still clearly a prominent position. The land slopes down from the Peach Orchard to the Trostle Farm and to the low ground 500 yards to your left. Yet the elevation where you are now standing is at eye level with the Peach Orchard. This shows the Peach Orchard could provide a plausible artillery platform from which to put fire onto the Union line where you are now standing.

Note also that you cannot see the ground beyond the Peach Orchard. You can see the treetops of Warfield Ridge but you cannot see what might be happening on the ground there. It means Union commanders at this position in 1863 could not see beyond the Peach Orchard and so they could not see what Confederate units might be forming battle lines or moving across it.

The infantry fighting on July 2nd came to an end at your location and to your left and right. At your front left, Col. George Willard's New York brigade repulsed the attack of Barksdale's Brigade once and for all. In front of you, the suicidal charge of the 1st Minnesota infantry brought a halt to Brig. Gen. Cadmus Wilcox's Alabama Brigade. And to your right, Brig. Gen. Ambrose Wright's Georgia Brigade struck Cemetery Ridge but had to withdraw for lack of support.

More Detail: Lee filed three reports of the battle, two in July and the last one in January 1864. It was in his January report that the general would say explicitly and for the first time that his priority accomplishment on July 2nd had been to secure the Peach Orchard as an artillery position for continued operations.

Look to your left.

The field leading from Cemetery Ridge to the Trostle Farm was the ground across which Willard led his regiments in the counter-charge that stopped the attack of Barksdale's Brigade.

Colville

Look to your front.

With the fighting over, Willard's brigade headed back to its original position at the north end of Cemetery Ridge.

In good light and in the right season, you may be able to pick out a dark tablet monument 350 yards down the slope beyond two fence lines. You might guess this marker has to have something to do with the 1st Minnesota infantry and its famous charge down this exact slope. It does not. It marks the spot where an artillery shell from the Peach Orchard hit Willard in the head and killed him as he rode back toward Cemetery Hill.

For most historians, this slope of ground is more important as the site of the famous charge of the 1st Minnesota infantry.

You can see the Klingel Farm and the Emmitsburg Road beyond the thicket. Wilcox's Alabama Brigade, moving in tandem with Barksdale's Brigade, reformed its lines to the right of the Klingel Farm and proceeded down the slope toward the thicket in front of you.

The section of Cemetery Ridge where you are presently located was wide open. The four brigades that had been in position here were redeployed to the Wheatfield a mile to your left and quite far out of sight. The 1st Minnesota infantry was the only unit in position here available to stop Wilcox. They had been assigned to this ridge by themselves to protect a U.S. Army artillery battery that was booming away to your right.

The regiment charged down this hill in a drama that befits another tour and a whole book of its own (an excellent one being Richard Moe's "Last Full Measure" in 2001). We will note just two points. First, two of the most well-documented Minnesota casualties, Col. William Colville and Pvt. Isaac Taylor, were both hit not by Alabama bullets but by fragments of shells fired from Alexander's guns in the Peach Orchard. Second, Colville would say nightfall saved his men from complete massacre – indicating the attack did indeed occur at the very end of the day.

Look to your right.

The wooded thicket along Hancock Avenue by the intersection with Pleasanton Avenue was occupied by the large 19th Maine infantry regiment. Their oblique fire saved many Minnesotans. Beyond their position, Brig. Gen. Ambrose Wright's Georgia Brigade crossed the Emmitsburg Road and hit the Union line, capturing a Union battery. How far the Georgians got remains an open question but they did eventually withdraw in the dark for lack of support.

With these repulses – Barksdale to your left, Wilcox to your front, Wright's Brigade to your right – this day of death and horror came to a close. Gunners in both Alexander's battalion and on McGilvery's Plum Run Line said they fired in the dark by aiming at the muzzle sparks from the opposing cannon. Everybody finally quit for the night and a stunned, mournful silence set down over a landscape of carnage.

Driving Tour Stop #19 (FINAL): The Upper Angle

- **ON THIS LEG OF THE DRIVE:** You will pass through and behind the main Union infantry line that would confront Pickett's Charge on July 3rd.
- **GO:** Drive up Hancock Avenue past the Pennsylvania Memorial on your right, passing the intersection with Pleasanton Avenue. Continue on Hancock Avenue past the cluster of Union monuments. Pass the U.S. Regulars Monument, a large obelisk on a square base, at mile point 0.2. Pass the Copse of Trees and the Highwater Mark Memorial on your left, pass the first marked parking spaces for NPS Auto Tour Stop #15 "The Angle" at mile point 0.3, continue to the part of the road where the stone wall bends north and continues close on your left, parallel with the road.
- **STOP:** park at the first available spaces on the right. Cross the street and go over to the wall.
- **LOOK:** Survey the battlefield landscape in front of you leading down to the Emmitsburg Road and left toward the red buildings of the Codori Farm in a thicket of trees. Follow the Emmitsburg Road south with your eye.

Quick Summary: You have reached the ground of Pickett's Charge, the climax of the Battle of Gettysburg. It includes the so-called "copse of trees" to your left. It includes the bend in the wall 80 yards in front of you at "The Angle" where the largest number of Confederates broke through. And it includes the wall at your feet that follows the upper ridge-line to the white barn of the Brian Farm to your far right.

Cemetery Hill was Lee's ultimate goal, the "decisive point," in his July 2nd attack. It is to your right rear.

Pickett's Charge on July 3rd was a disaster, and since it was the climax, it means the whole battle was a disaster for the Army of Northern Virginia.

Far to your front left, follow the Emmitsburg Road with your eye to find the prominent white-log house of the Klingel Farm. Beyond that is the red Sherfy barn and beyond that the Peach Orchard itself. This vantage illustrates that a force advancing from the Peach Orchard on July 2nd simply had too much distance to cover to ever reach the "decisive point," Cemetery Hill.

Meade

More Detail: The legendary Robert E. Lee had been forced to remain on the offensive constantly for the three days of the battle. The offensive posture suited Lee's nature. Yet at Gettysburg his impulse to remain on the attack had led him and his men to their defeat. It lost them the battle and from an entirely military perspective it lost them the war.

Consider the events of the battle you have been reviewing for the past 20 miles of your tour.

The famous charge and its repulse, the options and choices made by Lee and Meade, the errors their lieutenants did or did not make, have all been topics of discussion in the taverns and parlors of Gettysburg amid coffee, beer, wine, brandy and cigars, not to mention the interminable quarreling on Facebook.

If only Ewell had attacked Cemetery Hill. If only Longstreet had moved more quickly. If only Pender had not been wounded. If only the light howitzers Alexander was counting on for Pickett's Charge had not been moved. If only Major Gen. Jeb Stuart's cavalry division had reported sooner.

Weekend historians and armchair generals spend countless hours in such weeds. But these deep debates tend to feed an illusion. There emerges this notion that a small change in events might have changed the outcome – a Civil War version of the "butterfly effect."

Enough is enough.

Since the open-book monument in front of the trees to your left is often referred to as the "High Water Mark," perhaps the U.S. Navy can be of assistance. The Navy in the 1960s developed a command principle with the acronym K.I.S.S. – Keep It Simple, Stupid.

So here it is, keeping it simple.

1. Losing a battle is a mistake
2. Lee lost the Battle of Gettysburg.
3. Thus Lee made a mistake.

So where was this mistake?

The impulsive answer, considering where you are standing, is that his mistake was to order Pickett's Charge. Resist that impulse for a moment. Consider the possible answers to this one question: What was Lee's alternative to the infantry charge on July 3rd?

Was it plausible for Lee to withdraw from the battlefield after two days of fighting? No. Lee had advanced so far into Union territory yet the historic victory eluded him. He had adequate combat power to press the fight for a third day. Withdrawal was out of the question.

Was it plausible for Lee to adopt a defensive strategy on July 3rd and wait for the Union to attack him? Again, no.

Meade had drawn Lee into battle under conditions in which Lee was required to maintain the offensive or risk the survival of his army. Lee could not loosen his offensive grip for fear that Meade, if given adequate time and space, would send Union divisions down from the heights to smash Lee's flank.

Was it plausible for Lee to attack someplace else on July 3rd? No. Where?

In the morning, Lee had expected Longstreet to resume his attack from the Peach Orchard area with Pickett's fresh division in support or in the lead. Longstreet did not. Lee rode out to investigate the problem. Longstreet was able to convince him that his First Corps divisions had been all fought out on July 2nd, that the Union flank had been massively reinforced while his force was depleted, and that the divisions of McLaws and Law were the only protection for Lee's own right flank on Seminary Ridge. Lee saw it Longstreet's way. Lee amended his attack plan, shifting the focus of the July 3rd assault to the center of Cemetery Ridge.

Turn around to contemplate the equestrian statue honoring Meade.

The proud figure of Meade positioned on his faithful horse, Old Baldy, represents victory. His image as rendered by Henry Kirk Bush-Brown is that of an erect, solid, resolute commander. It is not a bombastic image of a general rushing to fight or cheering his men into an attack. Rather, Meade had set the strong position, he had set the battle with a successful concentration of his divisions and he had proceeded to win the battle by compelling Lee to attack him fruitlessly while Meade shifted forces to parry the blows, to protect his supply roads and to hold the high ground.

(By sheer chance and serendipity, the Meade equestrian, erected 1896, and the Lee equestrian, erected 1917, have the opposing commanders gazing at each other across that deadly mile.)

Many authors on the battle or on the entire Civil War find great fault with Meade. They claim he was tepid in his pursuit of Lee back to Virginia. They point out that Meade failed to bring Lee to battle in the ensuing six months of 1863 leading into winter. They then expand this critique to conclude Meade must have been lackluster at Gettysburg, that he had won a victory by luck and had little to do with the successful outcome. It is a case of inverted logic.

Indeed, Meade's career in command of the Army of the Potomac beyond July 4th can be said to have lacked a vigorous offensive strategy – a failing that would go uncorrected until the transfer of Gen. Ulysses S. Grant to the Eastern Theater in March 1864. But any such failing by Meade, as it were, ought not to be used to cast aspersion on his career from June 28th to July 3rd. In these six days he grasped quickly a complex situation, he maneuvered brilliantly, he kept his options open and his corps in close support of one another, and he compelled Lee to be drawn into his greatest defeat of the American Civil War.

Turn back toward the Emmitsburg Road and look to your right front.

Modern development and tree clusters obscure your view of the Lutheran Seminary. From this vantage point you can see the modern "Battle Theater" building on the Emmitsburg Road and the array of flags around it. The Lutheran Seminary is one mile beyond directly in line with that point.

Driving Tour Stop #1 – "The Lutheran Seminary" is where Lee formulated and explained his attack plan for July 2nd centered on the Peach Orchard. The goal was to roll up the Union flank and eventually to smash Cemetery Hill, the core point of the Union defense. It is Jomini's "decisive point." As you have seen from your present vantage, the Peach Orchard was simply too far away. Lee was asking too much of the men he thought were "invincible." It was this failure on July 2rd and its outcome – not Pickett's Charge – that lost him the Battle of Gettysburg.

MONUMENT PHOTOS without captions

PAGE 55: Texas Memorial
PAGE 59: 141st Pa.
PAGE 60: 1st Pa. Battery C
PAGE 63: 7th NJ
PAGE 66: 5TH United States Battery I
PAGE 67: 1st Minn. with detail from Pa. Memorial
PAGE 70: Meade Equestrian

Bonus Tour: Cycling

The "Cycling Tour" has its own directions designed to spare riders difficulties they would encounter on the Driving Tour or Hiking Tour. It features 18 stops over 10 miles and will use the narratives from the other tours. The tour avoids the top of Cemetery Hill and the Round Tops yet it still traverses rolling country where brief hills and steep ridges are an integral part of the battlefield terrain and so unavoidable. Riders can shorten the trip by three miles at Cycling Tour Stop #2 by skipping ahead to Cycling Tour Stop #8. Or riders with more ambition can add two miles of scenic country byways between Cycling Tour Stop #3 and Cycling Tour Stop #5.

Either way, the tour begins at the Lutheran Seminary, where Gen. Robert E. Lee made the decision to mass his assault force at the Peach Orchard in order to strike the Union left. The tour then directs you to follow the route of the Confederate divisions on their march to the battlefield. Many of the historic roads for this marching are still in place with limited traffic as they pass lovely fields and pastures. But these farm roads can be narrow and they lack good shoulders so caution and mirrors are necessary.

The Cycling Tour will bring you into Gettysburg National Military Park to vantage points selected to build a deeper appreciation for the importance of the Peach Orchard in the fighting on July 2nd. Most of the park tour roads are one-way. National Park Service rules allow bicycles to ignore the one-way directions and to go in the opposite direction against car and bus traffic but it is not recommended. Cyclists who go against traffic are taking a risk because most motorists will not be expecting it.

Cycling Tour Stop #1: The Lutheran Seminary

GO: From the National Park Service Visitor Center, leave by the Taneytown Road exit (west) and turn right onto Taneytown Road. Take your first left into the National Cemetery parking area. Ride through the parking area and follow the tour road down to the Emmitsburg Road. Turn right onto Emmitsburg Road, go three blocks and turn left onto Queen Street at the gas station and the Gettysburg Diorama Center. Continue to the "T" intersection with Long Lane and turn right. Continue straight at the Stop sign onto Franklin Street and turn left at High Street. Continue to the "T" intersection with Howard Avenue and turn right. Cross Middle Street (Fairfield Road) and turn left onto Springs Avenue. This will bring you to a brief but steep climb up onto Seminary Ridge. (It is better to attack the climb on a quiet backstreet than on busy Fairfield Road or Chambersburg Pike.)

Cyclists who begin their tour from downtown should go west on Middle Street from the Adams County Courthouse for five blocks to reach Howard Avenue. Turn right onto Howard Avenue and then left onto Springs Avenue for the climb up Seminary Ridge.

Cyclists who begin their tour from the circle in the center of town can choose to go west on Chambersburg Street. It is the shortest route but can involve more traffic, pedestrians and opening car doors. At the third traffic light, the intersection with West Street, continue absolutely straight onto Springs Avenue. (The prevailing traffic bears right into Buford Avenue/Chambersburg Pike.)

Once up the ridge at Springs Avenue, turn right onto Seminary Ridge Avenue and go 150 yards to reach the historic Lutheran Seminary building, four stories tall with the green-domed cupola on top. It is presently the home of the Seminary Ridge Museum. Dismount and walk your bikes around the left (north) side to the "back" of the building. This leads to the ground-floor museum entrance. Consult the narrative from "Driving Tour Stop #1: Lutheran Seminary" on page 36.

Cycling Tour Stop #2: Thompson House/ "Lee's Headquarters"

GO: Leaving the Seminary Ridge Museum, turn right onto Seminary Ridge Avenue and ride north to the Stop sign on Buford Avenue/Chambersburg Pike. The Thompson house is directly in front of you across the street. Dismount and carefully walk your bike across the street and enter the parking lot for "Lee's Headquarters," the Mary Thompson House. Consult the narrative under "Driving Tour Stop #2: Lee's Headquarters" on page 38.

SHORTER TOUR: You can shorten the 10-mile tour by three miles if you skip Cycling Tour Stops #3 through #7 (Longstreet's countermarch) and proceed directly to Cycling Tour Stop #8: "Warfield Ridge War Dept. Observation Tower." Leaving the Lee's Headquarters parking lot, carefully walk your bike back across Chambersburg Pike onto Seminary Ridge Avenue. Ride straight past Cycling Tour Stop #1 "The Lutheran Seminary" and continue straight to the traffic light. Cross Fairfield Road onto West Confederate Avenue. Go straight for two miles to the Stop sign at Millerstown Road. Cross Millerstown Road and proceed to the War Dept. Observation Tower on Warfield Ridge. Dismount by the pair of large black 20 lb. Parrott Rifles positioned near the base of the tower. To the left of the guns you will see a notch in the stone wall leading to a short path. Follow it out into the field. Consult the narrative under "Driving Tour Stop #8: Warfield Ridge War Dept. Observation Tower" on page 49.

Cycling Tour Stop #3: Herr Ridge

GO: Leaving the Lee's Headquarters (Thompson House) parking lot, make a right onto Chambersburg Pike and head west. Pass the West End Guide Station on McPherson's Ridge and cross Willoughby Run, the national park boundary. At the top of the ridge you will reach historic Herr Tavern and the Herr Ridge Road with a pair of traffic lights. Using EXTREME CAUTION, turn left at the second light onto historic Herr Ridge Road. Stop along the right shoulder. Consult the narrative under "Driving Tour Stop #3: Herr Ridge" on page 40.

Cycling Tour Stop #4: Forney Farm Lane Location

GO: Ride straight on Herr Ridge Road away from the Chambersburg Pike behind you. Cross the intersection with Old Mill Road. Pass in sequence on your left Woodhaven Drive, Bittern Drive and finally Kestrel Drive. You will come to a private black mailbox on the right, beyond which there is a good grass and gravel shoulder where you can stop and pull off. Consult the narrative under "Driving Tour Stop #4: Forney Farm Lane Location" on page 41.

ALTERNATIVE SCENIC TOUR: You may take a more scenic route past beautiful fields and pastures. It will add two miles to your trip and will bring you to a short but difficult hill. To follow this addition to the tour, proceed from Cycling Tour Stop #3 "Herr Ridge," skip Cycling Tour Stop #4 and follow these directions instead to proceed directly to Cycling Tour Stop #5 "Black Horse Tavern/Fairfield Road."

From Cycling Tour Stop #3 "Herr Ridge," follow Herr Ridge Road to the intersection with Old Mill Road. Turn right onto Old Mill Road. At the Stop sign, turn left onto Knoxlyn Road. At the next intersection, turn left onto Blackhorse Tavern Road and follow it 1.7 miles to the Fairfield Road. Dismount and walk to your left to the iron signs along the shoulder in front of the Black Horse Tavern. Consult the narrative under "Driving Tour Stop #5: Black Horse Tavern/Fairfield Road" on page 42. Then proceed by following the directions for "Cycling Tour Stop #6: Breams' Hill Countermarch Point."

Cycling Tour Stop #5: Black Horse Tavern/Fairfield Road

GO: Continue on Herr Ridge Road past two Fairplay Road intersections until you reach Breams Hill Road. Turn right on Breams Hill Road and follow to the Fairfield Road. Dismount and walk to your right to the iron signs along the shoulder in front of the Black Horse Tavern. Consult the narrative under "Driving Tour Stop #5: Black Horse Tavern/Fairfield Road" on page 42.

Cycling Tour Stop #6: Breams' Hill Countermarch Point

GO: Carefully cross the Fairfield Road and follow the Blackhorse Tavern Road south past the intersection with Plank Road. Follow Blackhorse Tavern Road southeast up the steep, brief climb over Breams Ridge. Cross the top of the ridge to come into full view of the open fields and distant tree line denoting Warfield Ridge. You might be able to see the top of the War Dept. observation tower on Warfield Ridge. If so, pull onto the grassy shoulder just beyond the private shed on the right. If trees block your view of the observation tower, move up to the white gravel driveway on your left until you can see the tower. Consult the narrative under "Driving Tour Stop #6: Breams' Hill Countermarch Point" on page 46.

Cycling Tour Stop #7: Willoughby Run Road/Blackhorse Tavern Road

GO: Continue east on Blackhorse Tavern Road a half a mile to the intersection on the left with Willoughby Run Road. Stop on the right. Consult the narrative under "Driving Tour Stop #7: Willoughby Run Road/Blackhorse Tavern Road" on page 47.

Cycling Tour Stop #8: Warfield Ridge War Dept. Observation Tower

GO: Continue east and south on Blackhorse Tavern Road across Willoughby Run and to the Millerstown Road. Turn left onto Millerstown Road and ride past the white-fence driveway entrance to Eisenhower Farm. Continue up the slope, approaching Warfield Ridge from the rear. At the Stop sign, turn right onto West Confederate Avenue. Proceed to the War Dept. observation tower on Warfield Ridge. Dismount by the pair of large black 20 lb. Parrott Rifles positioned near the base of the tower. To the left of the guns you will see a notch in the stone wall leading to a short path. Follow it out into the field. Consult the narrative under "Driving Tour Stop #8: Warfield Ridge War Dept. Observation Tower" on page 49.

Cycling Tour Stop #9: Snyder Farm

GO: Ride south on West Confederate Avenue away from the Millerstown Road. Pass the Georgia and South Carolina memorials on your right. Pass beyond the end of the Tower Woods on your left. Pass the iron sign identifying Biesecker Woods. When you reach the gap in the stone wall on your left for a dirt trail adjacent to the Snyder House, dismount and park your bike. Walk up the dirt horse trail path about 100 feet, roughly halfway to the Emmitsburg Road. Consult the narrative under "Hiking Tour Stop #5: Snyder Farm" on page 19 and/or "Driving Tour Stop #10: Snyder Farm" on page 54.

Cycling Tour Stop #10: Texas Memorial on Warfield Ridge

GO: Return to your bike and ride south on West Confederate Avenue past the Arkansas Memorial on your right to the Stop sign on Emmitsburg Road. Cross the Emmitsburg Road with caution – the speed limit here is 40 mph and many drivers exceed the limit. Ride up to the first artillery battery you see on the left. Dismount and walk over to the guns. Consult the narrative under "Driving Tour Stop #11: Texas Memorial on Warfield Ridge" on page 55.

Cycling Tour Stop #11: Rose Farm Lane

GO: Ride back to the Emmitsburg Road and turn right (north) up the Emmitsburg Road. Stop when you reach the Rose Farm Lane on your right. Consult the narrative under "Hiking Tour Stop #4: Rose Farm Lane" on page 17.

Cycling Tour Stop #12: 2nd New Hampshire Monument, Southwest Corner Peach Orchard

GO: Continue north up the Emmitsburg Road until you reach the first edge of the Peach Orchard at the intersection with Birney Lane on your right. Dismount and walk your bike to the pyramid monument to the 2nd New Hampshire infantry. Consult the narrative under "Walking Tour Stop #3: 2nd New Hampshire Monument, Southwest Corner" on page 7.

Cycling Tour Stop #13: 1st New York Battery G, Northwest Corner Peach Orchard

GO: Return to the Emmitsburg Road and ride up the shoulder about 100 yards to the next corner at the intersection with Wheatfield Road. Dismount and walk over to the pair of green Napoleon cannon and the monument between them. You may also choose to walk your bike to that corner along the western edge of the Peach Orchard. (DO NOT ride your bike on the grass or through the trees. Off-road cycling in the park is prohibited.) Consult the narrative under "Walking Tour Stop #4: 1st New York Battery G, Northwest Corner" on page 9.

Cycling Tour Stop #14: 141st Pennsylvania Monument, Northeast Corner Peach Orchard

GO: Return to the Emmitsburg Road and turn right onto the Wheatfield Road. Ride east along the northern edge of the Peach Orchard past the monuments for 1st Pennsylvania Battery C and Battery F. Dismount at the corner with the asphalt steps at the thick obelisk monument for the 141st Pennsylvania infantry. Consult the narrative under "Walking Tour Stop #5: 141st Pennsylvania Monument, Northeast Corner" on page 11.

Cycling Tour Stop #15: John Weikert Lane

GO: Continue riding east on Wheatfield Road past the Wheatfield, Ayres Avenue and across Plum Run. Stop where the dirt lane from the John Weikert Farm comes in from the left. Consult the narrative under "Hiking Tour Stop #13: John Weikert Lane" on page 30.

Cycling Tour Stop #16: Weikert Woods

GO: Continue riding east on Wheatfield Road up the slope to the Stop sign. Turn left onto Sedgwick Avenue to ride north. Pass the Sedgwick equestrian monument on your right and continue north for about 400 yards – a quarter mile. Pause along the stone wall on the left side of the road. You may be able to see a dirt horse trail about 200 yards across the rocky field on your left (west.) Consult the narrative under "Hiking Tour Stop #12: Weikert Woods" on page 28. Bear in mind the narrative is written for someone on the other side of this field on that horse trail. Begin your reading by facing south in the direction from which you just rode.

Cycling Tour Stop #17: 1st Minnesota Memorial

GO: Continue riding north on Sedgwick Avenue past the intersection with United States Avenue from the left, at which point the road changes names to Hancock Avenue. Ride past the line of batteries and monuments along your left until you reach the Minnesota Memorial about a quarter of a mile ahead on your left. It is a tall rectangular pedestal with a statue of the soldier advancing with his bayonet out front. Stop beyond the monument. Consult the narrative under "Driving Tour Stop #18: 1st Minnesota Memorial" on page 67.

Cycling Tour Stop #18 (FINAL): The Upper Angle

GO: Continue riding north on Hancock Avenue past the Pennsylvania Memorial on your right and the intersection with Pleasanton Avenue. Continue on Hancock Avenue past the cluster of Union monuments. Pass the U.S. Regulars Monument, a large obelisk on a square base on your left. Pass the Copse of Trees and the open-book High Water Mark Memorial on your left. Pass the first marked parking spaces for NPS Auto Tour Stop #15 "The Angle." Ride up to the part of the road where the stone wall bends north and continues close on your left, parallel with the road. Stop and dismount. Consult the narrative under "Driving Tour Stop #19: The Upper Angle" on page 68.

GO: If you started your Cycling Tour at the Visitors Center of Gettysburg National Military Park, you can return there by continuing along Hancock Avenue to the Stop sign. Turn right and pass through the National Cemetery parking lot to the Stop sign at Taneytown Road. Turn right onto Taneytown Road and make your second left into the Visitors Center.

If you started your Cycling Tour in the middle of Gettysburg, you can return there by continuing along Hancock Avenue to the Stop sign. Turn left and take the park road down the hill to the Stop sign at the Emmitsburg Road. Turn right onto the Emmitsburg Road and continue straight to the first traffic light. Turn left onto Washington Street. A ride of seven blocks will bring you to Chambersburg Street at the second traffic light. A right turn and a one-block ride will bring you to the traffic circle at the center of town but it might be best to walk your bike that last bit.

Bonus Tour: Couch

The "Couch Tour" skips the concerns about turns, stops, traffic and the need to follow a sequence of events dictated by the landscape. Here the armchair generals are free to fly about the field with their fingers, traveling directly to the most interesting and provocative points.

As with the Driving Tour, the Couch Tour reader is directed to begin at the Lutheran Seminary, where Confederate Gen. Robert E. Lee made the decision to mass his assault force on July 2nd at the Peach Orchard in order to drive in the Union Army's left.

The reader is then directed toward specific "tour-stop" narratives in keeping with five general themes. The reader should skip the GO-STOP-LOOK directions and go directly to the "Quick Summary" and "More Detail" narratives.

The narrative readings as presented in the Couch Tour are not in the same sequence as the field tours and so some officers will need to be identified on first reference (although good armchair generals already know the names of all these guys.)

Command Decisions: A battle lost, a battle won.

The largest ideas of the tour (and accompanying volume "Peach Orchard: The Battle of Gettysburg Reconsidered") are incorporated here, with discussion of how Gen. Robert E. Lee made the error of committing to a battle he would lose. Also discussed is the point, too often neglected, of how Gen. George Gordon Meade won the thing.

Be advised the narrative under Driving Tour Stop #5 will introduce Union Brig. Gen. **Andrew** Humphreys, who is marching two brigades northwestward up the Blackhorse Tavern Road en route to danger. Under Driving Tour Stop #18 we are introduced to New York Col. George Willard, albeit briefly, seeing as his time left on this earth is brief. Also mentioned here is Confederate Brig. Gen. William Barksdale, whose time left on this earth is equally brief.

- ☛ **Driving Tour Stop #1: Lutheran Seminary** (page 36)
- ☛ **Driving Tour Stop #2: Lee's Headquarters** (page 38)
- ☛ **Driving Tour Stop #5: Black Horse Tavern/Fairfield Road** (page 42)
- ☛ **Driving Tour Stop #18: 1st Minnesota Memorial** (page 67)
- ☛ **Driving Tour Stop #19: The Upper Angle** (page 68)

Command decisions: The Congressman is confused.

A secondary aspect of the "Command Decisions" theme involves a character who in no way appears secondary: Major Gen. Daniel Edgar Sickles, the ambitious, disgraced New York City political hack. Sickles recruits a brigade and is rewarded with a general's commission. He proceeds to squirm and schmooze his way upward to corps command, an example of a defective promotions system that on July 2nd would cost many lives.

- ☛ **Driving Tour Stop #10: Snyder Farm** (page 54)
- ☛ **Hiking Tour Stop #4: Rose Farm Lane** (page 17)
- ☛ **Hiking Tour Stop #12: Weikert Woods** (page 28)

Command Decisions: "It was unmistakably anything but the flank…"

As misdirected as Sickles, if not moreso, was the embattled, tormented Major Gen. Lafayette McLaws. Set on the march along a supposed concealed route to the Peach Orchard, he discovers one mistake when cresting Breams Hill and a second mistake when arriving on Warfield Ridge. McLaws' plans for the day (as well as his outlook and career) would never recover. The quote in the title is from 21st Mississippi Col. **Benjamin** Humphreys (as opposed to Union Gen. Andrew Humphreys. Union = Andrew; Confederate = Benjamin.)

- **Driving Tour Stop #6: Breams Hill Countermarch Point** (page 46)
- **Driving Tour Stop #7: Willoughby Run Rd./Blackhorse Tavern Rd.** (page 47)
- **Driving Tour Stop #8: Warfield Ridge War Dept. Observation Tower** (page 49)
- **Driving Tour Stop #11: Texas Memorial on Warfield Ridge** (page 55)

Infantry: Fighting for the Peach Orchard line.

Command controversies aside, the fighting along the Peach Orchard line, once joined, was fierce and hot as Confederate columns, heavy with ammunition and peak morale, smashed Union regiments determined to hold the home soil of Pennsylvania. It was in this fight that the Confederate forces won the ground and lost the war.

- **Walking Tour Stop #1: Peach Orchard Salient Wayside Sign.** (page 2)
- **Walking Tour Stop #3: 2nd New Hampshire Monument.** (page 7)
- **Hiking Tour Stop #4: Rose Farm Lane** (page 17)
- **Hiking Tour Stop #9: Sherfy Farm** (page 24)
- **Walking Tour Stop #5: 141st Pennsylvania Monument** (page 11)
- **Driving Tour Stop #15: NPS Auto Tour Stop 10 Peach Orchard** (page 62)

Artillery: Gunfight at the Peach Orchard

A thrilling element of the Peach Orchard fighting on July 2nd was the close, direct-fire action of so many gun batteries under command on both sides by the best artillery officers of the war, Confederate Col. E. Porter Alexander and Union Lt. Col. Freeman McGilvery. Time passed as the Confederate infantry divisions redeployed and rearranged their order of attack. It was during this time that increasing numbers of guns hammered away at one another at remarkably close range. At the climax of this "gunfight," an astounding total of 86 guns were engaged on both sides, all of this occurring before the infantry stepped off into their attack.

- **Hiking Tour Stop #7: Louisiana Memorial** (page 21)
- **Driving Tour Stop #8: Warfield Ridge War Dept. Observation Tower** (page 49)
- **Hiking Tour Stop #6: South End Tower Woods** (page 20)
- **Hiking Tour Stop #5: Snyder Farm** (page 19)
- **Walking Tour Stop #2: "Twin Threes" 3rd Mich./3rd Me. Flank Markers.** (page 5)
- **Walking Tour Stop #4: 1st New York Battery G** (page 9)
- **Driving Tour Stop #15: NPS Auto Tour Stop 10 Peach Orchard** (page 62)
- **Walking Tour Stop #5: 141st Pennsylvania Monument** (page 11)
- **Hiking Tour Stop #10: Trostle Farm** (page 25)
- **Driving Tour Stop #17** (page 65) **or Hiking Tour Stop #11: Plum Run Line** (p. 26)

Credit for the victory… must go to the General who devised and executed the plan that forced the enemy to make a mistake.

Donald Kagan

"Thucydides"

Photo credits:

All officer & period images public domain from Library of Congress, National Archive and Gettysburg National Military Park
All maps, color monument and color landscape photos by Ralph Siegel…

except as follows –

✓Author photo Preface page ii (left) Bobby Housch **GETTYSBURGDAILY.COM**
✓Author photo Preface page ii (right) PCN-Pennsylvania Cable Network
✓ 3rd Maine monument page 7 by Darah Gardner facebook **Djgardner Photography / FINEARTAMERICA.COM/**PROFILES/DARAH-GARDNER.HTML
✓9th Mass. Battery Wheatfield Road monument page 26 by Taryn Kerper facebook **GETTYSBURG PICTURES**
✓Emmitsburg Road landscape (view from Warfield Tower) page 49 by Darah Gardner
✓Peach Orchard colorized page 58 by Taryn Kerper
✓Author photos Back Page (top) Jerry Hahn / (down right) PCN-Pennsylvania Cable Network

GALE HOUSE PUBLISHING, TRENTON NJ

Key to quotes on Back Cover

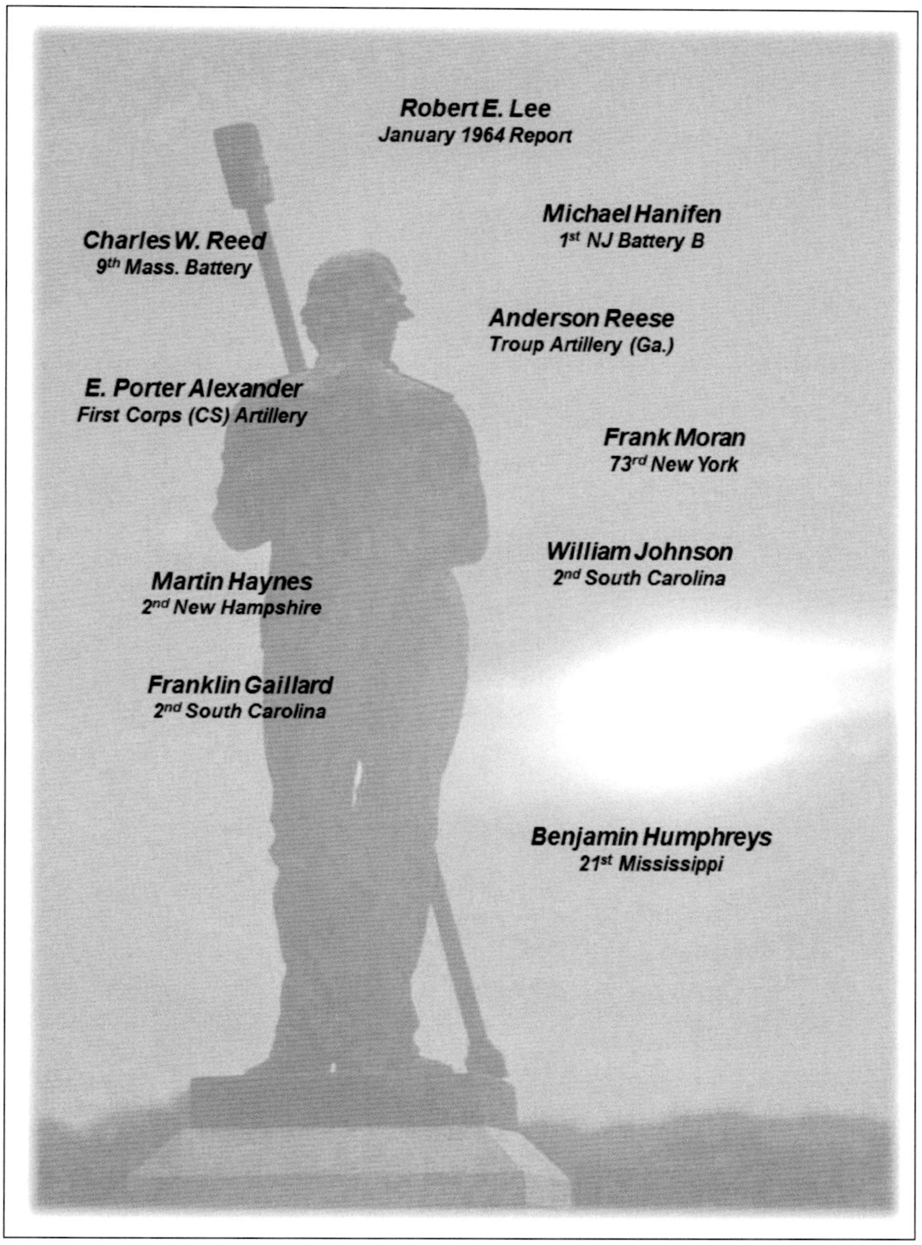

Made in United States
Orlando, FL
04 November 2026

90200942R00052